# VEDIC MATHEMATICS
# NEW HORIZONS
# **INITIAL LESSONS**

Published by:
**Lotus Press, Publishers & Distributors**

# VEDIC MATHEMATICS
# NEW HORIZONS
# INITIAL LESSONS

DR. S.K. KAPOOR

*Ved Rattan*

4735/22, Prakash Deep Building,
Ansari Road, Daryaganj,
New Delhi - 110002.

**Lotus Press: Publishers & Distributors**
Unit No. 220, 2nd Floor, 4735/22, Prakash Deep Building,
Ansari Road, Daryaganj, New Delhi - 110002.
Ph.: 32903912, 23280047 • E-mail: lotus_press@sify.com
www.lotuspress.co.in

**Vedic Mathmatics New Horizons Initial Lessons**



ISBN: 978-81-8382-299-2 (HB)



*Published by:* **Lotus Press Publishers & Distributors,** New Delhi
*Printed at:* **Concept Imprint**, Delhi

# Foreward

Swami Bharti Krishna Tirtha Ji Maharaj well demonstrated potentialities of Ganita Sutras. The illustrative applications of Ganita Sutras worked out by Swami Ji, attracted the attention of the world of mathematics. But the Discipline again has started having signs of stagnation for want of further clues. As it appears that the scholars are getting confined, to the domain of the illustrative demonstrations given by Swami Ji.

Being clueless, as to further applications of Ganita Sutras, the subject has not entered the main stream of knowledge. Now with these two volumes of initial and advance lessons of Dr. S. K. Kapoor, new hope of revival of the Ancient Discipline has become lively. Inspired by the works of Dr. Kapoor, we, the i360 Staffing and Training Solutions Pvt. Ltd., company in collaboration with the Unicon Trust and Paramount Knowledge Pvt. Ltd. Company are planning to take up the Project: 'Vedic Mathematics' for its dissemination and research.

We shall be requesting Dr. S. K. Kapoor to help us structure Vedic Mathematics Teachers Courses and soon we would like to take up these courses in the right earnest.

I am sure that these LESSONS, as the Title conveys shall be leading the Students, Teachers, Scholars and Institutions dedicated for 'Dissemination and Research' in the field of Vedic Mathematics, to New Horizons of the Discipline.

**(Mrs. Sonia Nagpal)**
Chairperson
i360 Staffing and Training
Solutions Pvt. Ltd.
New Delhi.

# Preface

Vedic Mathematics is an Ancient Discipline. It has its own postulates and processing systems. As such its attainments are distinctively of different values and virtues.

Being a Discipline, it has to be learnt step-by-step. It is with this aim that the present INITIAL LESSONS have been planned. These are based upon the sequence and features of Ganita Sutras and Upsutras.

Depending upon the background of previous exposure, one may, during first reading, skip over the sequence of these Initial Lessons. It is during the second reading that one may again re-charter the chase steps for one self and finally one may follow the sequence of Lesson 1 to Lesson 70.

Here teachers have a big role to play, to see that the young minds do not have mental blocks which may, in any way cause any conceptual comprehension difficulties for the Advance lessons to follow in the second book.

Naturally, it is to depend upon the intensity of urge to know more about Vedic Mathematics and the efforts to learn this Discipline, which may be the determining factors of the success. However, the ultimate factor which is to matter more would be the privilege one shall be having is to be the guidance of the teacher.

As such, like other systems of knowledge, the system of the Discipline of the knowledge as well, at least at the initial stages need be approached as a class-room subject. It is to be sequentially learnt. As such these lessons have been designed within an aim that those who wish to be Vedic Mathematics teacher, they can 'learn and teach by being through these initial lessons', in the sequence and order of these lessons. With this in mind these initial lessons are being published. With the hope that these may help the seekers of knowledge of Vedic Mathematical domain in particular.

**Dr. S. K. Kapoor**

# Contents

# 1
# VEDIC KNOWLEDGE

## Vedas

1. '*Ved*' (वेद) has a root '*vid*' (विद्) of two fold values, 'knowledge' and 'to know'. Vedas as such are the reservoir of whole range of pure and applied values of knowledge.
2. Originally there was a wholesome 'Ved', for the whole range of knowledge. Subsequently it was reorganized by Maharishi Ved Vyas as four vedas, namely, Rigved, Yajurved, Samved and Athravved. Further these vedas are of 21, 101, 1000 and 9 branches respectively. Each branch has a distinct Samhita, Brahmana, Arnik and Upnishad. And this way there are 1131 × 4 = 4524 basic vedic scripture. Still further, each ved has a distinct Upved.
3. The four upveds are Ayurved, Dhanurved, Ghandaravved and Sathapatyaved. Of these, Ayurved, as the name well indicates, is all about 'age', and hence about 'health concerns for whole span of life'. Dhanurved is all about weapons and war. Ghandaraved is all about blissful sounds, while Sathapatyaved is all about 'Sathapana'/manifestations and there mathematics, sciences and technologies.
4. Sathapatya ved, as such is the source reservoir of the values of 'Vedic mathematics, science and technology',

which in a way, covers mathematics, sciences and whole range of technologies of Vedic order.

5. Vedic mathematics is one specific value of vedic knowledge. It is the core of the Discipline of Vedic Mathematics, science and technology. The Sathapatya Upved, being the source reservoir of the discipline of Vedic Mathematics, science and technology, the seat of whole range of features of Vedic mathematics, as such are blissfully lively in the scriptures of Sathapatya Ved.
6. Many many values of Vedic knowledge have remained dormant for centuries. Credit goes to Swami Bharti Krishna Tirthaji Maharaj to focus the attention of present generation world of mathematics about the potentialities and applied values of Ganita Sutras traced from the Parshista (appendix portion) of Atharavved.
7. The text of Ganita Sutras (as well as of its Upsutras) consist of just 16 Sutras and 13 Upsutras availing in all 520 letters, and as such these 16 Sutras (and 13 Upsutras) can be memorized with a few recitations thereof and with this memorization of the Sutras, whole of the rest of the exercise of availing the values and working rules of these Sutra and Upsutra would become just a 'mental exercise', because of which Vedic mathematics is designated and is known as 'mental mathematics'.
8. However, the basic reason for acceptance of Ganita Sutras mathematics has mental mathematics is inherently flowing out of the sequential organization of these hymns which sequentially unfold their values parallel to the intelligence field within the human mind grows and unfolds itself.
9. This being the blissful feature of mathematics of Ganita Sutras, as such solemn responsibility is cast upon the teachers of Vedic mathematics to very gently lead the young minds to the natural flow path of INTELLIGENCE.

❑ ❑ ❑

# 2

# GANITA SUTRAS TEXT

| Sutra No. | **Sutra text** (Original in Sanskrit, with Roman script and with simple English Rendering) |
|---|---|
| 01 | **ॐ। एकाधिकेन पूर्वेण।**<br>*Om / Ekadhiken Purvena.*<br>Om । By One More than One before. |
| 02 | **एकाधिकेन निखिलं नवतश्चरमं दशतः।**<br>*Nikhilam Navatascramam Dasatah.*<br>All from 9 and the last from ten. |
| 03 | **ऊर्ध्वतिर्यग्भ्याम्।**<br>*Urdhva tiryagbhyam.*<br>Vertically and crosswise. |
| 04 | **परावर्त्य योजयेत्।**<br>*Paravartya Yojayet.*<br>Transpose and Apply. |
| 05 | **शून्यं साम्समुच्चये।**<br>*Sunyam Samyasamuccaye.*<br>If the samuccaya is the same it is Zero. |
| 06 | **(आनुरूप्ये) शून्मन्यत्।**<br>*(Anurupye) Sunyamanyat*<br>If one is in Ration the others is Zero. |

07 संकलव्यवकलनाभ्याम्।

*Sankalana-vyavakalanbhyam.*

By addition and by subtraction.

08 पूरणापूरणाभ्याम्।

*Puranapuranabhyan.*

By the completion or non-completion.

09 चलनकलनाभ्याम्।

*Calana-kalanabhyam*

Differentiation Calculus.

10 यावदूनम्।

*Yavadunam.*

By the Deficiency.

11 व्यष्टिसमिष्टिः।

*Vyastisamastih.*

Specific and General.

12 शेषाण्यङ्केन चरमेण।

*SesnyankenaCaramena.*

The Remainder by the last digit.

13 सोपान्त्यव्दमन्तयम्।

*Sopantyadvyamantyam.*

The ultimate and twice the penultimate.

14 एकन्यूनेन पूर्वेण।

*Ekanyunena Purvena.*

By One less than the One Before.

15 गुणितसमुच्चयः।

*Gunitasamuccayah.*

The product of the Sum.

16 गुणकसमुच्चयः।

*Gunaksamuccayah*

All the Multipliers.

# 3

# GANITA UPSUTRAS TEXT

| Upsutra No. | **Upsutra text** (Original in Sanskrit, with Roman script and with simple English Rendering) |
|---|---|
| 01 | **आनुरूप्ये ।**<br>Anurupyena.<br>Proportionately. |
| 02 | **शिष्यते शेषसंज्ञः ।**<br>*Sisyate Sesasamjnah.*<br>That remains is remainder. |
| 03 | **आधमाधेनान्त्यमन्त्येन ।**<br>*Adyamadyen Antyamantyen.*<br>First with first Last with last. |
| 04 | **केवलैः सप्तकं गुण्यात् ।**<br>*Kevalaih Saptakam Gunyat.*<br>Only Seven as multiplicand. |
| 05 | **वेष्टनम् ।**<br>*Vestanam*<br>Osculators. |
| 06 | **यावदूनं तावदूनम् ।**<br>*Yavadunam Tavadunam.* |

That twice This twice.

07 यावदूनं तावदूनीकृत्य वर्गं च योजयेत्।

*Yavadunam Tavadunikrtya Varganca Yogayet.*

That twice This twice Square and add.

08 अन्त्ययोर्दशकेऽपि।

*Antyayordasake'pi.*

Ends to sum as ten.

09 अन्त्ययोरेव।

*Antyayoreva.*

Ends to be in ratio.

10 समुच्चयगुणितः।

*Samuccayagunitah.*

Samuchya as product.

11 लोपना स्थापनाभ्याम्।

*Lopana Sthapananabhyam.*

That missing to be established.

12 विलोकनम्।

*Vilokanam.*

By observation.

13 गुणितसमुच्चयः समुच्चयगुणितः

*GunitaSamuccaya Samuccayagunitah.*

Product samuchya Samuchya Product.

❑ ❑ ❑

# 4

# COMPLETE SCRIPTURE

## 1. Introductory

1. Complete Scripture means a single range from beginning to end.
2. Beginning of the complete scripture shall has its definite source reservoir.
3. The end of the complete scripture as well has its definite end fruit values and virtues to be of the order of the start with source reservoir.
4. Vedic scriptures accept Om (ॐ) as its source reservoir.
5. The end fruit values and virtues of the Vedic scripture are of the order of Parnava (प्रणव:), the synonym of Om (ॐ).

## 2. Ganita Sutras

6. Ganita Sutras (including Ganita upsutras) is a complete Vedic scripture.
7. As such, it has sole syllable Om (ॐ) as its source reservoir.
8. The end fruit values and virtues of this scripture are of the un-manifest range parallel to three syllables (i) प्र (ii) ण (iii) व: ; the third of these being of the pair of features of the formats of single and double flow streams.

9. This as such, makes a system, whose end fruit values and virtues, which are of un-manifest feature, are to be made to flow as of a pair of streams.
10. This as such would help supplement three additional upsutras to make Sutras and Upsutras being of parallel flow streams, complementary and supplementary of each other at every step.

## 3. Flow Range of Ganita Sutras and Upsutras

11. Flow range of Ganita Sutras and Upsutras is parallel to the composition of Ganita Sutras and Upsutras availing sole syllable Om (ॐ) and 519 letters making it beginning to the end of the manifested part of flow range being of 520 steps.
12. The end fruit virtues and values of this manifest flow range shall be of 3 + 2 + 3 = 8 un-manifest flow steps.
13. These un-manifest flow range of eight steps shall be permitting two fold approaches to its middle as from either end and thereby it would split and reorganize 8 = 3 + 5 = 5 + 3.
14. It shall be a split parallel to the organization of the transcendental (5-space) domain as of a solid order setup. The transcendental (5-space) domain as it emerges at the origin seat of creator space (4-space), acquires a pair of features folds parallel to the pair of dimensions of the spatial order of the creator space (4-space).
15. It is because of it that the organization 3 + 2 + 3 is symmetrical for its middle.
16. However its split it in asymmetrical set-up of {(3 + 2), 3}.
17. Simultaneously it shall be making available the parallel but of opposite orientation set-up {3, (2 + 3)}.
18. With it the pair of splits shall be [{(3 + 2), 3}, {3, (2 + 3)}].
19. As such, the structural range of Ganita Sutras shall be of the format of hyper cube 5, which has creative boundary of ten components (hyper cube 4s) and as within each

hyper cube, cube shall be acquiring an additional edge, i.e. 13th edge, along each of the four dimensions, so the swapping of the creative domain shall be of the feature of $13 \times 4 = 52$ components, which for all the ten creative boundary components of transcendental (5-space) domain shall be $52 \times 10 = 520$ parallel to 520 letters range of the Ganita Sutras and Upsutras text.

20. Such external coverage in terms of 520 letters range, together with the un-manifest values and virtues range of Asht Prakrati being of self-referral dimensional order shall be parallel to the self-referral origin fold of the transcendental (5-space) domain, and with it this scripture of Ganita Sutras shall be of the order and values of the transcendental (5-space) domain itself.

## 4. Formulation 'गणित सूत्राणि:' Ganita Sutrani'

21. The formulation 'गणित सूत्राणि:' is of six syllables range, namely

    '(i) ग (ii) णि (iii) त (iv) सू (v) त्रा (vi) णि:'

22. The pair of Sub formulations of this formulation are:

    '(i) गणित (ii) सूत्राणि'

23. Both these sub formulations are of three syllables each.

24. The first sub formulation '(i) गणित' further admits split as a pair of sub formulations namely (i) गण् (ii) इत.

25. The formulation 'गण' means army/group/set of entities. It leads to counting enumeration components in setting.

26. The formulation 'इत' means to remove and to set it into flow.

27. Ganita गणित as such would mean the discipline of enumeration of components in setting to be removed and to be set into a flow.

28. The formulation Sutrani means to have spectrum mode flow streams with its inner states set as surfaces like that

of liquefied solid set into flow as states layers of currents of stream.

29. The formulation Ganita as Discipline of pure values and the formulation Sutrani as the discipline of applied values makes 'Ganita Sutrani' as the complete discipline of pure and applied values.
30. The presiding deity of complete discipline of pure and applied values of group of entities is Lord Ganesh, the Lord of group of entities.

❑ ❑ ❑

# 5

# GANITA SUTRA-1

# 'एकाधिकेन पूर्वेण' / EKADHIKEN PURVENA

# BY ONE MORE THAN ONE BEFORE

## 1. General

1. Read and recite the text of the Sutra.
2. Sit comfortably and think about the working rule of Ganita Sutra-1 as is expressed in simple English rendering: 'By one more than before'.

## 2. Arithmetic

3. Express this working rule in Arithmetic language as follows:

   1. 1 + 1, (1 + 1) + 1, {(1 + 1) + 1} + 1, ..................

4. Express above steps, by accepting and designating them as counts: '1 = one, 1 + 1 = two, (1 + 1) + 1 = three, {(1 + 1) + 1} + 1 = four' and so on.
5. Express above steps in their distinct symbolic forms:

   '1 = one = 1

   1 + 1 = two = 2

   (1 + 1) + 1 = three = 3

   {(1 + 1) + 1} + 1 = four = 4'

   and so on.

6. Accept (1, 2, 3, 4, ..................) as counting numbers.

7. Accept by definition, counting numbers as natural numbers.

## 3. Geometry

8. Express the working rule steps (1, 1 + 1, (1 + 1) + 1, {(1 + 1) + 1} + 1, ..................) and so on, in geometric language as '1 = one = Length

   1 + 1 = two = Length and Breadth

   (1 + 1) + 1 = three = Length, Breadth and Height

   {(1 + 1) + 1} + 1 = four = Length, Breadth, Height and Time and so on.

9. Express further these features as

   '1 = one = Length = Line

   1 + 1 = two = Length and Breadth = Surface

   (1 + 1) + 1 = three = Length, Breadth and Height = Solid

   {(1 + 1) + 1} + 1 = four = Length, Breadth, Height and Time = Hyper solid - 4 and so on.

10. Still further as

    '1 = one = 1-space

    1 + 1 = two = 2-space

    (1 + 1) + 1 = 3-space

    {(1 + 1) + 1} + 1 = 4-space and so on.

11. Still further expresses

    '1 = one = 1 – space body; interval

    1 + 1 = two = 2 – space body; square

    (1 + 1) + 1 = 3 – space body; cube

    {(1 + 1) + 1} + 1 = 4 – space body; Hyper cube 4

    and so on.

## 4. Algebra

12. Express the working rule steps (1, 1 + 1, (1 + 1) + 1, {(1 + 1) + 1} + 1, ..................) and so on in geometric language

as one variable (x), two variables (x, y) and three variables (x, y, z) and so on.

13. Further express as

    '1 = one, one degree $x^1$

    1 + 1 = two, two degrees $x^2$

    (1 + 1) + 1 = three, three degrees $x^3$

    {(1 + 1) + 1} + 1 = four, four degrees $x^4$

    and so on.

14. Still further express as

    '1 = first degree equation

    $ax + b = 0$

    1 + 1 = second degree equation

    $ax^2 + bx + c = 0$

    (1 + 1) + 1 = third degree equation

    $ax^3 + bx^2 + cx^1 + d = 0$

    {(1 + 1) + 1} + 1 = fourth degree equation

    $ax^4 + bx^3 + cx^2 + dx + e = 0$

    and so on.

❑ ❑ ❑

# 6

# PROCESSING APPROACH

## 1. Vedic Systems

1. Vedic systems are as per the Vedic knowledge values. These systems process and workout the values of knowledge as organized as Vedic scripture. The organization formats beneath this scripture on the knowledge of the scripture organized on these formats run parallel to each other. This may be viewed as knowledge and organization of knowledge aspects. Two fold basic feature at work in the parallel flow of knowledge and organization of knowledge in Vedic scripture respectively along the text of the scripture and the formats beneath thereto. The established processing processes are designated and known as Sankhiya Nishtha and Yoga Nishtha.

## 2. Sankhiya Nishtha and Yoga Nishtha

2. Sankhiya Nishtha on ultimate analysis presumes the existence of geometric formats and avails artifices of numbers for chase of the values of the geometric formats. On the other hand the Yoga Nishtha on its ultimate analysis presumes the existence of artifices of numbers and avails dimensional frames for chase of the values of the geometric formats.

## 3. Artifices and Dimensional Frames

3. This way the pair of mathematical entities emerging as of presumed existence are, first the dimensional frames and secondly the artifices of numbers as dimensional frames help chase values of artifices of numbers and on the other hand as the artifices of numbers help chase the geometric formats as such there these features of the complementary and supplementary of each other and also as these being parallel to the steps, as such there relations intersee and sequential progressions thereof make out a three fold processing approaches, which in that sequence become the discipline and also their basis being 'arithmetic', geometric, and algebraic values and formulations.

## 4. Approach to Ganita Sutras

4. Parallel to about three fold approaching approach, when Ganita Sutras would be approached, these shall be unfolding and reading the Discipline of Ankganit/ arithmeitc, Rekhaganit/ geometry and Beejganit/algebra.
5. Accordingly, first of all as the Sankhiya Nishtha approach. The values of the discipline of Ank Ganit (arithmetic) to be sequentially reached at by following the working rules (of upsutras) in the sequence and order of their texts.
6. Thereafter, parallel to Yoga Nishtha approach to Ganita Sutras, the values of discipline of geometry may be unfolded in the sequence and order of Ganita Sutras (and Ganita upsutras).
7. With it, a phase and stage would arise to work out the relationships of artifices of numbers and dimensional frames as values of the discipline or algebra/beejganit.

## 5. Text of Sutras

8. The text of Sutras, as has reached us is a Devnagri script. As such to be fully familiarized with it, and to have intimate friendly bonds with these hymns (sutras). It

would be advisable to first of all get intimately familiarized with the Devnagri alphabet in Devnagri script itself in which the hymns composition are with us.

9. However till one is well acquainted with the Devnagri script, one can start working with the Sutras text in Roman script, as that at initial stages the working rules of the Sutras expressed in simple English rendering may serve the purpose of initial stages steps of mathematics being learnt.
10. One shall pronounce and recite the Sutras number of times so that one starts acquiring familiarity and intimacy with these hymns.
11. A time spent in reciting the Sutras will go a long way to acquire insight about the working rules of the Sutras.

## 6. Individual Letters of the Text

12. Every individual letter of the text significantly matters. Moreover letter by letter chase of the text of the Sutra shall be helping to have utmost intimacy with the Sutra and it shall be providing the deepest insight about the values and virtues of the Sutras and applications of their working rules.
13. As the super structure of the discipline of mathematics is to be built upon the foundations of the Sutras, as such the teachers are under solemn responsibility to help young minds to acquire desire degree of intimacy with the text of the Sutras to ensure perfection of their insight and for imbibing of the values of the working rules of the Sutras.

❑ ❑ ❑

# 7

# ARITHMETIC OPERATIONS

## Outline Steps

1. Addition

   Basic rule: Sutra 1 'One more than before'

2. Subtraction

   Basic rule: Sutra 14 'one less than before'

3. Multiplication

   (*i*) As repeated addition (at base)

   (*ii*) As repeated addition (at index)

A. By the rule of Nikhilam Sutra

   (*a*) By deficiency from the base

   (*b*) By deficiency from sub base

      (*i*) Multiple of base

      (*ii*) Part of base (sub base)

B. By Udhrav Sutra

   (*c*) Digit wise

      (*i*) Single digit multiplication

      (*ii*) Double digit multiplication

      (*iii*) Triple digit multiplication

      (*iv*) Higher digit multiplication

   (*d*) Geometric format

      (*i*) for the beginning and last digits

C. By Upsutra – 3

(*ii*) For the middle part

(*e*) Multiplication of numbers of value 1 for all digits

(*i*) 1 × 1 =1

(*ii*) 11 × 11 =121

(*iii*) 111 × 111 = 12321

(*iv*) 1111 × 1111 = 1234321

And so on.....

4. Division operation
   By repeated subtraction
   (*i*) At base
   (*a*) By Paravartya Sutra and argumentation
   (*b*) Straight division of placement values for algebraic expressions formats for single variable
   (*ii*) At the index
5. Squaring and cubing
   (*a*) Squaring as multiplication by Urdharav Sutra
   (*b*) Cubing as repeated multiplication by Urdharav Sutra
   (*c*) Argumentative straight squaring
   (*d*) Argumentative straight cubing
   (*e*) Tests of perfect squares
   (*f*) Tests for perfect cubes
6. Square roots and cube roots
   (A) By properties of last digits of perfect squares
   (*a*) By properties of the last two digits of perfect squares
   (*b*) By properties of the last three digits of perfect squares
   (B) By properties of last digits of perfect cubes
   (*a*) By properties of the last two digits of perfect cubes
   (*b*) By properties of the last three digits of perfect cubes
7. Decimal expressions for fractions
   (*a*) Expressions of finite places of decimal representation
   (*b*) Infinite range expressions of decimal representations

(*i*) Recurring ranges

(*ii*) Non recurring ranges

## Fractions

7. Values of fractions of the time 1/n, n as a natural number

   1/1 = 1
   1/2 = .5
   1/3 = .33 ...............
   1/4 = 1/2 × 2
   1/5 = .2
   1/6 = 1/2 × 3
   1/7 = .(142857) (142857) ...............
   1/8 = 1/2 × 4
   1/9 = 1/3 × 3
   1/10 = 1/2 × 5
   1/11 = .(01) (01) ...............
   1/12 = 1/2 × 6
   And so on

8. Conversion of last digit of odd (other than five) denominator of the fraction as 9

   (*i*) 9 × 1 = 9, as such if last digit is one, the numerator and denominator can be multiplied by 9

   (*ii*) 3 × 3 = 9, as such if last digit is three, the numerator and denominator can be multiplied by 3

   (*iii*) 7 × 7 = 49, as such if last digit is seven, the numerator and denominator can be multiplied by 7

9. **Oscillators:** For fractions of recurring decimal expressions, definite oscillators corresponding to recurring features of the decimal expressions can be found by the features of last digit of the denominators.

10. **Auxiliary factors:** Because of the osculating features of numbers, corresponding auxiliary fractions can be determinant which will help reach at the values of the

decimal expression by the process of straight multiplication and division steps in terms of the determinant auxiliary factors.

11. **HCF, LCM and Solutions of simple quadratic and cubic equations:** By the process of decomposition of the middle terms of products.

    The middle-term of the products as a structure permitting split as per the contributions of the crosswise multiplication and summation as per the Udharav Sutra format. This as such, in its reverse steps from the obtained product back to the Udharav steps shall be argumentatively leading to the factors in terms of which HCF, LCM and factorization of algebraic expressions can be reached at.

12. **Transition from linear to angular approach:** Sutra–2 rule 'all from nine and last from ten' shall be providing the steps for a shift from the linear to angular approach to reach from an line to surface by reaching from a line to an angle, measure in terms of the format for total angle at a point parallel to the circumference of the circle with the given point as its centre.

13. It would be parallel to the place value approach of numbers for a numbers line. The ten place value system as such shall be providing a measure of 360 degree for the total angle of the surface at a point of the surface. A change to another place value system, say to six place value system, as such shall be leading to '320' degree value measure for the total angle at a point approached by the six place value system. This 320 value measure of six place value system, as such would be equal to $3 \times 6 \times 6 + 2 \times 6 + 0 = 120$ value of ten place value system. Like that different place value system shall be working out different values measures for corresponding transitions from linear to angular set ups.

❑ ❑ ❑

# 8

# REVISITING GANITA SUTRA-1

### 1. Sutra: Sanskrit Text:

'एकाधिकेन पूर्वेण:'

Text in Roman Script: 'Ekadhiken Purvena'

Simple English rendering 'one more than before'

### 2. Construction of Whole Numbers

'1' is 'one more than the previous (0)'.

'2' is 'one more than the previous (1)'.

'3' is 'one more than the previous (2)'.

And so on.

### 3. Construction of Dimensional Frame

| | |
|---|---|
| 'Point' | has no length and as such is of 'zero' dimension/axis. |
| 'Line' | has length and as such it has one dimension than before that is of (of point). |
| 'Surface' | has area and as such it has two dimensions/axes that is, one more than the previous (one dimension of length). |
| 'Solid' | has volumme and as such it has three dimensions/axes, that is one more than the previous (two dimension of surface/area' |

## 4. Construction of Axes / Dimensional Frames

One axis — It can be constructed with the help of ruler/scale.
It shall be of the features of the straight line.
It shall be representing the length of a line.

Two axes — It can be constructed with the help of ruler and a set square.
It shall be of the features of the pair of crossing lines.
It shall be representing as length and breadth of surface area.

Three axes — These three axes can be constructed as three axes emanating from the corner point of a cubeboid
Three axes shall be representing as length, breadth and height of solids.

## 5. Interval, Square and Cube

Interval — Interval is a geometric body accepting one axis/ one axis frame/one dimensional frame.
As such 'interval' would be a 1-space body.

Square — Square is a geometric body accepting a pair of axes/two axes frame/two dimensional frame.
As such square would be a 2-space body.

Cube — Cube is a geometric body accepting three axes/ three axes frame/three dimensional frame.
As such cube would be 3-space body.

## 6. Origin Point

Origin point — origin point formally may be defined and accepted as point where axes cross.
The corner points of the cube would be the origins of three axes frames.
The corner points of the square would be the origin points of two axes frames.
The end points of the interval would be the origin points of the one axis frame.

❑ ❑ ❑

# 9

# REVISITING GANITA SUTRA-14

## 1. Sutra: Sanskrit Text:

Text in Sanskrit 'एकान्यूनेन पूर्वेण:'

Text in Roman Script: 'Ekanuena Purvena'

Simple English rendering 'one less than before'

## 2. Direct Counting and Reverse Counting

The Sutra 'एकाधिकेन पूर्वेण:'/'Ekadhiken Purvena'/'one more than before' leads to a rule of direct counting/counting in ascending order leading to counts/set of counting numbers (1, 2, 3, 4, 5, 6, 7, 8, 9, ........).

This sutra 'एकान्यूनेन पूर्वेण:'/'Ekanuena Purvena'/'one less than before' leads to a rule of reverse counting/counting in descending order leadings to counts sequence (........100, 99, 98, 97, ........ 9, 8, 7, 6, ........).

As such, direct counting and reverse counting may be taken as a pair of orientations because of ascending order as well as because of descending order.

If ascending order is taken as orientation/sequential progression from east to west, the opposite orientation because of descending order may be taken as sequential progression from west to east.

## 3. Counting Line

The ascending order sequential progression of direct counts sequence (1, 2, 3, 4, 5, 6, 7, 8, 9, ........) may be taken as constituting a counting line running through the counts (points, of number values 1, 2, 3, 4, 5, 6, 7, 8, 9, ........) as under:

The descending order sequential progression of reverse counts sequence (100, 99, 98, 97, 96, ........) may be taken as constituting a counting line running through the counts (points, of number values 100, 99, 98, 97, 96 ........) as under:

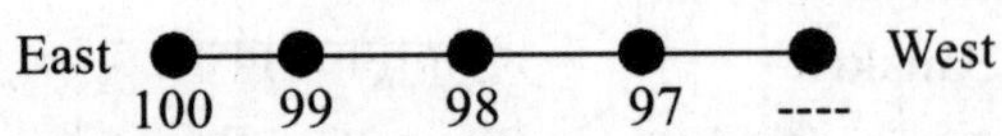


## 4. Positive and Negative Counts

We may formally accept the counts (1, 2, 3, 4, 5, 6, 7, 8, 9) as positive counts. (+1, +2, +3, +4, ........). Parallel to positive counts, we may construct the set of counts (/01, -2, -3, -4, ........) to be designated as negative counts.

## 5. Construction of Positive and Negative Counts

Positive counts are constructed by the Ekadhikena Sutra rule of one more than before as that (0 + 1 = + 1, +1 + 1 = + 2, + 2 + 1 = + 3, ........).

Negative counts are constructed by the Ekanuena Sutra rule of one less than before as that (0 – 1 = –1, –1 – 1 = –2, –2 – 1 = –3, ........).

## 6. Positive Counts as Half Line from West to East

The positive counts line, by definition we taken as a half line from west to east orientation as follows.

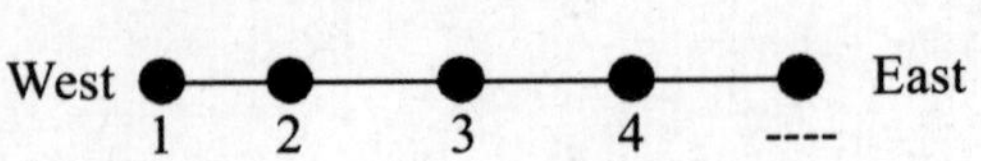

## 7. Negative Counts as Half Line from East to West

The negative counts line, by definition we taken as a half line from east to west orientation as follows:

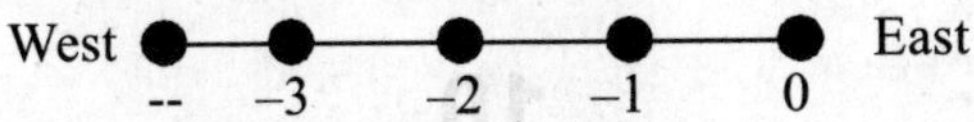


## 8. Full Counting Line

Positive counts half line from west to east and negative counts half line from east to west, together may be taken as here full counting line as of expression as follows:

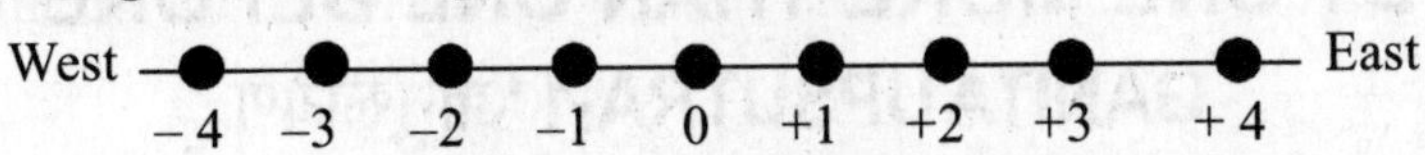


## 9. Positive negative orientation of a Line

It be taken by definition that west to east orientation of line is the positive orientation and like wise, parallel to it, east to west orientation of line to be taken as the negative orientation of line.

❑ ❑ ❑

# 10

# GANITA SUTRA-1
# 'एकाधिकेन पूर्वेण'/ EKADHIKEN PURVENA BY ONE MORE THAN ONE BEFORE'
## GANITA UPSUTRA-1 'आनुरूप्येण ANURUPYENA / FOLLOW THE FORM AS FRAMED/ SYMMETRY/PROPORTIONATELY'

### 1. General

1. Read and recite Sutra-1.
2. Read and recite Upsutra-1.
3. Read and recite Sutra-1 and Upsutra-1.
4. Think about the working rule of Sutra-1.
5. Think about the working rule of Upsutra-1.
6. Think about the working rule of Sutra-1 together with the working rule of Upsutra-1.

### 2. Direct and Reverse Counting

7. Count in increasing sequence as (1, 2, 3, 4, 5, ........).
8. Count in decreasing sequence as (........, 10, 9, 8, 7, 6, 5, ........).

### 3. Counting with One Jump

9. Count in increasing sequence with one jump at every step as (1, 3, 5, 7, 9, ........)

(0, 2, 4, 6, 8, ........)

10. Count in decreasing sequence with one jump at every step as (........, 10, 8, 6, 4, ........)

    (........, 11, 9, 7, 5, ........)

## 4. Counting with Two Jump

Count in increasing sequence with two jumps at every step as

(1, 4, 7, 10, 13, ........)

(0, 3, 6, 9, 12, ........)

11. Count in decreasing sequence with two jump at every step as

    (........, 20, 17, 14, 11, ........)

    (........, 21, 18, 15, 12, ........)

## 5. Positive and Negative Numerals

12. Positive numerals (1, 2, 3, 4, 5, 6, 7, 8, 9)

    Negative numerals (–1, –2, –3, –4, –5, –6, –7, –8, –9)

## 6. Special Symbols for Negative Numerals

13. ($\overline{1}$ = –1, $\overline{2}$ = –2, $\overline{3}$ = –3, $\overline{4}$ = –4, $\overline{5}$ = –5, $\overline{6}$ = –6, $\overline{7}$ = –7, $\overline{8}$ = –8, $\overline{9}$ = –9)

14. $\overline{0}$ = –0 =0

15. Negative numbers ($\overline{1}$, $\overline{2}$, $\overline{3}$, $\overline{4}$, $\overline{5}$, $\overline{6}$, $\overline{7}$, $\overline{8}$, $\overline{9}$, $\overline{10}$, $\overline{11}$, $\overline{12}$, $\overline{13}$, $\overline{14}$, $\overline{15}$, ...)

## 7. Parallel Lines

16. One shall, with the help of Ganita Sutra-1, reach at line of positive orientation.
17. Further with the help of Sutra-1 and Upsutra-1 reach at the line of negative orientation.
18. Still further to reach at pair of parts of a line as positive orientation of a line and negative orientation of line.
19. Still further to reach at a pair of parallel lines.

❑ ❑ ❑

# 11

# SUTRA-1 AND UPSUTRA-2 APPLICATIONS

## 1. Counting with Jumps

1. Counting with jumps is at the base of the tables.
2. Counting with one jump i.e. (0, 2, 4, 6, ........) is at the base of table-2.
3. Counting with two jump i.e. (0, 3, 6, 9, 12, ........) is at the base of table-3.
4. Counting with three jump i.e. (0, 4, 8, 12, ........) is at the base of table-4.
5. Like that all tables are just counting with appropriate jumps.

## 2. Vinculum Numerals

6. Parallel to the numerals (1, 2, 3, 4, 5, 6, 7, 8, 9) are the vinculum numerals ($\bar{1}$, $\bar{2}$, $\bar{3}$, $\bar{4}$, $\bar{5}$, $\bar{6}$, $\bar{7}$, $\bar{8}$, $\bar{9}$).
7. The numbers 1 to 100 can be written as ten rows and ten columns as under

| | | | | | | | | | |
|---|---|---|---|---|---|---|---|---|---|
| 01 | 02 | 03 | 04 | 05 | 06 | 07 | 08 | 09 | 10 |
| 11 | 12 | 13 | 14 | 15 | 16 | 17 | 18 | 19 | 20 |
| 21 | 22 | 23 | 24 | 25 | 26 | 27 | 28 | 29 | 30 |
| 31 | 32 | 33 | 34 | 35 | 36 | 37 | 38 | 39 | 40 |

| 41 | 42 | 43 | 44 | 45 | 46 | 47 | 48 | 49 | 50 |
|---|---|---|---|---|---|---|---|---|---|
| 51 | 52 | 53 | 54 | 55 | 56 | 57 | 58 | 59 | 60 |
| 61 | 62 | 63 | 64 | 65 | 66 | 67 | 68 | 69 | 70 |
| 71 | 72 | 73 | 74 | 75 | 76 | 77 | 78 | 79 | 80 |
| 81 | 82 | 83 | 84 | 85 | 86 | 87 | 88 | 89 | 90 |
| 91 | 92 | 93 | 94 | 95 | 96 | 97 | 98 | 99 | 100 |

8. The vinculum expression values for 1 to 100 may be tabulated as follows:

| 001 | 002 | 003 | 004 | 005 | $11\bar{4}$ | $01\bar{3}$ | $01\bar{2}$ | $01\bar{1}$ | 010 |
|---|---|---|---|---|---|---|---|---|---|
| 011 | 012 | 013 | 014 | 015 | $12\bar{4}$ | $02\bar{3}$ | $02\bar{2}$ | $02\bar{1}$ | 020 |
| 021 | 022 | 023 | 024 | 025 | $13\bar{4}$ | $03\bar{3}$ | $03\bar{2}$ | $03\bar{1}$ | 030 |
| 031 | 032 | 033 | 034 | 035 | $14\bar{4}$ | $04\bar{3}$ | $04\bar{2}$ | $04\bar{1}$ | 040 |
| 041 | 042 | 043 | 044 | 045 | $15\bar{4}$ | $05\bar{3}$ | $05\bar{2}$ | $05\bar{1}$ | 050 |
| 051 | 052 | 053 | 054 | 055 | $1\bar{4}4$ | $1\bar{4}3$ | $1\bar{4}2$ | $1\bar{4}1$ | $1\bar{4}0$ |
| $1\bar{4}1$ | $1\bar{4}2$ | $1\bar{4}3$ | $1\bar{4}4$ | $1\bar{4}5$ | $1\bar{3}4$ | $1\bar{3}3$ | $1\bar{3}2$ | $1\bar{3}1$ | $1\bar{3}0$ |
| $1\bar{3}1$ | $1\bar{3}2$ | $1\bar{3}3$ | $1\bar{3}4$ | $1\bar{3}5$ | $1\bar{2}4$ | $1\bar{2}3$ | $1\bar{2}2$ | $1\bar{2}1$ | $1\bar{2}0$ |
| $1\bar{2}1$ | $1\bar{2}2$ | $1\bar{2}3$ | $1\bar{2}4$ | $1\bar{2}5$ | $1\bar{1}4$ | $1\bar{1}3$ | $1\bar{1}2$ | $1\bar{1}1$ | $1\bar{1}0$ |
| $1\bar{1}1$ | $1\bar{1}2$ | $1\bar{1}3$ | $1\bar{1}4$ | $1\bar{1}5$ | $10\bar{4}$ | $10\bar{3}$ | $10\bar{2}$ | $10\bar{1}$ | 100 |

## 3. Steps for Writing Tables

9. Outline of the steps for writing a table are as under:

Step 0: Write the number.

Step 1: Write vinculum equivalent of the number.

Step 2: Put the vinculum expression digit-wise in column.

Step 3: Write first term of the table as the number itself to be written digitwise in columns.

Step 4: Sequentially write the values in each column by adding the vinculum digits values.

Step 5: Table can be written up till any number of steps.

**Note:** However when the values becomes negative, the same can be made zero or positive by having a carry backward/ forward from/to the previous column,

10. Illustration: Table 19

Step 0: 19

Step 1: $2\bar{1}$

Step 2:

| Col. 2 | Col. 1 |
|---|---|
| 2 | $\bar{1}$ |

Step 3:

| | Columns | Col. 2 | Col. 1 |
|---|---|---|---|
| | Rule | 2 | $\bar{1}$ |
| Step 1 | 1 × 19 | 1 | 9 |

Step 4:

| | Columns | Col. 2 | Col. 1 | Value |
|---|---|---|---|---|
| | Rule | 2 | $\bar{1}$ | |
| Step 1 | 1 × 19 | 1 | 9 | 19 |
| Step 2 | 2 × 19 | 1+2=3 | 9–1=8 | 38 |
| Step 3 | 3 × 19 | 3+2=5 | 8–1=7 | 57 |
| Step 4 | 4 × 19 | 5+2=7 | 7–1=6 | 76 |
| Step 5 | 5 × 19 | | | |

11. Illustration Table 67

Step 0: 67

Step 1: $7\bar{3}$

Step 2:

| Col. 3 | Col. 2 | Col. 1 |
|---|---|---|
| 1 | $\bar{3}$ | $\bar{3}$ |

Step 3:

| | Columns | Col. 3 | Col. 2 | Col. 1 |
|---|---|---|---|---|
| | Rule | 1 | $\bar{3}$ | $\bar{3}$ |
| Step 1 | 1 × 67 | 0 | 6 | 7 |

Step 4:

| | Columns | Col. 3 | Col. 2 | Col. 1 | Value |
|---|---|---|---|---|---|
| | Rule | 1 | $\bar{3}$ | $\bar{3}$ | |
| Step 1 | 1 × 67 | 0 | 6 | 7 | 67 |
| Step 2 | 2 × 67 | 0+1=1 | 6–3=3 | 7–3=4 | 134 |
| Step 3 | 3 × 67 | 1+1=2 | 3–3=0 | 4–3=1 | 201 |
| Step 4 | 4 × 67 | **2+1=3**<br>**3-1=2** | **0–3=–3**<br>**–3–1= –4**<br>**10–4=6** | **1–3=–2**<br>**10–2=8** | **268** |
| **Step 5** | **5 × 67** | **2+1=3** | **6–3=3** | **8–3=5** | **335** |

**12. One can perfect one's comprehensions skills of writing tables by taking different exercises for the numbers 1 to 100.**

**13. Higher tables to any number value for any number of steps as well can be sequentially written on the above lines.**

**14. Firstly one shall write the tables from 1 to 10 and then from 11 to 100 and then from 101 to 1000 and so on from 1001 to 10000 and so on.**

❑ ❑ ❑

# 12

# GANITA SUTRA-2

# 'निखिलं नवतश्चरमं दशतः / NIKHILAM NAVATASCRAMAM DASATAH/ALL FROM NINE AND THE LAST FROM TEN'

## 1. General

1. Read and recite the text of the Sutra.
2. Sit comfortably and think about the working rule of Ganita Sutra-2 as is expressed in simple English rendering : 'All from 9 and the last from ten'.

## 2. Arithmetic

3. The working rule distinguishes between the roles of 9 and 10.
4. Infact, it as such separates the roles of the range '1 to 9' and that of '10'.
5. Numbers '1 to 9' are designated as the numerals of ten place value system.
6. This as such 'includes 0' in the range of '1 to 9' and makes it the range '0 to 9'.
7. Places of the place value format acquire sequential values in the order $10^0$, $10^1$, $10^2$ and so on.
8. This, in a way, with application of the rules of Ganita Sutra-1 and the Upsutra-1 shifts the sequential steps

increase as one more than before from 'base' to the index of '10'.

9. One may have a pause and have a fresh look upon the range '0, 1, 2, 3, 4, 5, 6, 7, 8, 9'.
10. One shall be noticing the feature of this range being as it to be having a single digit expression for its all the ten sequential steps.
11. A step ahead the range of the numbers '10 to 99' shall be a range of double digit numbers.
12. A step ahead the range of numbers 100 to 999 is going to be a range of triple digit numbers.
13. Here again, one may have a pause and see as that the single digits range '0 to 9' may permit expression as of double digits expressions as '00, 01, 02, 03, 04, 05, 06, 07, 08, 09'.
14. As such 100 double digit numbers range 00 to 99 shall be permitting organization as 10 × 10 matrix format of 10 rows and 10 columns.
15. The double digits range '01 to 99' shall be permitting its re-organization along the 9 × 11 matrix format as follows.

| | | | | | | | | |
|---|---|---|---|---|---|---|---|---|
| 01 | 02 | 03 | 04 | 05 | 06 | 07 | 08 | 09 |
| 10 | 11 | 12 | 13 | 14 | 15 | 16 | 17 | 18 |
| 19 | 20 | 21 | 22 | 23 | 24 | 25 | 26 | 27 |
| 28 | 29 | 30 | 31 | 32 | 33 | 34 | 35 | 36 |
| 37 | 38 | 39 | 40 | 41 | 42 | 43 | 44 | 45 |
| 46 | 47 | 48 | 49 | 50 | 51 | 52 | 53 | 54 |
| 55 | 56 | 57 | 58 | 59 | 60 | 61 | 62 | 63 |
| 64 | 65 | 66 | 67 | 68 | 69 | 70 | 71 | 72 |
| 73 | 74 | 75 | 76 | 77 | 78 | 79 | 80 | 81 |
| 82 | 83 | 84 | 85 | 86 | 87 | 88 | 89 | 90 |
| 91 | 92 | 93 | 94 | 95 | 96 | 97 | 98 | 99 |

❑ ❑ ❑

# 13

# GANITA SUTRA-2 APPLICATIONS FEATURES OF 9 × 11 ORGANIZATION FORMAT

## 1. Double Digit Numbers 01 to 99

1. Double digit numbers 01 to 99 permit re-organization along 9 × 11 matrix format as under:

| | | | | | | | | |
|---|---|---|---|---|---|---|---|---|
| 01 | 02 | 03 | 04 | 05 | 06 | 07 | 08 | 09 |
| 10 | 11 | 12 | 13 | 14 | 15 | 16 | 17 | 18 |
| 19 | 20 | 21 | 22 | 23 | 24 | 25 | 26 | 27 |
| 28 | 29 | 30 | 31 | 32 | 33 | 34 | 35 | 36 |
| 37 | 38 | 39 | 40 | 41 | 42 | 43 | 44 | 45 |
| 46 | 47 | 48 | 49 | 50 | 51 | 52 | 53 | 54 |
| 55 | 56 | 57 | 58 | 59 | 60 | 61 | 62 | 63 |
| 64 | 65 | 66 | 67 | 68 | 69 | 70 | 71 | 72 |
| 73 | 74 | 75 | 76 | 77 | 78 | 79 | 80 | 81 |
| 82 | 83 | 84 | 85 | 86 | 87 | 88 | 89 | 90 |
| 91 | 92 | 93 | 94 | 95 | 96 | 97 | 98 | 99 |

2. This re-organization along 9 × 11 matrix format splits the format itself into upper part and lower part. The separation numbers line runs through the numbers 10, 20, 30, 40, 50, 60, 70, 80, 90.

## 2. Upper Part

3. Let us have a fresh look at the upper part of above 9 × 11 organization format of double digit numbers being of the format set-up.

4. Let us again have a pause and have a fresh look at the above set up of the double digit numbers of the upper part and take a note of the mirror line running through the numbers 11, 22, 33 and 44.

5. This mirror line would permit extension on its either side separating numbers of the pair (01, 10) and (45, 54).

6. The features of these two pairs of numbers are that numbers of each pair swap digits of each other, as much as that the digits of 01 and 10 swap places and further the digits of (45 and 54) as well as swap places.

7. Such pairs of numbers which swap places for their digits as in case of (01, 10) and (45, 54) are defined and designated as reflection pairs of numbers.

8. One may again have a pause and have a fresh look at the set up of the numbers of upper half of 9 × 11 matrix format and see as that the numbers one side of the middle mirror line being constituting reflection pairs with the numbers on the other side of this numbers mirror.

9. It shall be a blissful exercise to enlist the total reflection pair of numbers of the upper part of 9 / 11 matrix format.

## 3. Lower Part

10. Let us have a fresh look at the lower part of above 9 × 11 organization format of double digit numbers being of the format set-up.

11. Let us again have a pause and have a fresh look at the above set up of the double digit numbers of the lower

part and take a note of the mirror line running through the numbers 55, 66, 77, 88 and 99.

12. One may again have a pause and have a fresh look at the set-up of the numbers of lower half of 9 × 11 matrix format and see as that the numbers one side of the middle mirror line being constituting reflection pairs with the numbers on the other side of this numbers mirror.
13. It shall be a blissful exercise to enlist the total reflection pair of numbers of the lower part of 9 / 11 matrix formats.

❑ ❑ ❑

# 14

# GANITA SUTRA-2 AND UPSUTRA-1 FEATURES OF 8 × 10 ORGANIZATION FORMAT

## 1. Nine Place Value System

1. With the help of the rule of 'following the form as framed/ symmetry/proportionately of Upsutra-1', and of the organization format of double digit numbers of ten place value system along the $9 \times 11 = 10 - 01 \times 10 + 01$ format, one may reach at the organization format for double digit of nine place value system $(9 - 1) \times (9+1) = 8 \times 10$ matrix format as follows:

| | | | | | | | |
|---|---|---|---|---|---|---|---|
| 01 | 02 | 03 | 04 | 05 | 06 | 07 | 08 |
| 10 | 11 | 12 | 13 | 14 | 15 | 16 | 17 |
| 18 | 20 | 21 | 22 | 23 | 24 | 25 | 26 |
| 27 | 28 | 30 | 31 | 32 | 33 | 34 | 35 |
| 36 | 37 | 38 | 40 | 41 | 42 | 43 | 44 |
| 45 | 46 | 47 | 48 | 50 | 51 | 52 | 53 |
| 54 | 55 | 56 | 57 | 58 | 60 | 61 | 62 |
| 63 | 64 | 65 | 66 | 67 | 68 | 70 | 71 |
| 72 | 73 | 74 | 75 | 76 | 77 | 78 | 80 |
| 81 | 82 | 83 | 84 | 85 | 86 | 87 | 88 |

## 2. Upper Part

2. One may have a pause here and have a fresh look at the above 80 double digit numbers of 9 place value system organized along 8 × 10 matrix format.
3. One may notice as that the numbers line (10, 20, 30, 40, 50, 60, 70, 80) splits the above format into two parts namely upper part and lower part.
4. The numbers above the numbers line including the numbers of the numbers line itself, namely (10, 20, 30, 40, 50, 60, 70, 80) constitute the upper part of the 8 ×10 matrix format.
5. One may again have a pause and have a fresh look at the numbers mirror (11, 22, 33, 44) permit extension towards one side and separates the pairs of numbers (01, 10) and makes it a reflection pair.
6. It would be a blissful exercise to enlist the reflection pairs of the upper part of above 8 × 10 matrix format.
7. The numbers (11, 22, 33, 44) may be designated by definition as the numbers of self-reflecting artifices pairs as both digits of each of these numbers (11, 22, 33, 44) are of equal values.

## 3. Lower Part

8. One may have a pause here and have a fresh look at the above 80 double digit numbers of 9 place value system organized along 8 × 10 matrix format.
9. One may again have a pause and have a fresh look at the numbers mirror (55, 66, 77, 88).
10. It would be a blissful exercise to enlist the reflection pairs of the lower part of above 8 × 10 matrix format.
11. The numbers (55, 66, 77, 88) may be designated by definition as the numbers of self reflecting artifices pairs as both digits of each of these numbers (55, 66, 77, 88) are of equal values.

❑ ❑ ❑

# 15

# SYMBOLS OF ARITHMETIC OPERATIONS

Arithmetic, geometry and algebra are three different groups of approach values of mathematics. Initially, these deserve to be learnt separately one sequence of learning is to start with arithmetic and then to shift to geometry and thereafter approach the algebra.

Ank (अंक) and Sankhya (संख्या) are two basic terms which may be familiarized first. At initial stage ank (अंक) may be taken as numerals 1 to 9 of 10 place value system and Sankhya (संख्या) to be taken as all numbers expressible in terms of the numerals, including the numerals themselves.

Of the numerals, the first numeral, namely 'one' deserves to be comprehended well and then with its help, as per the rule of Ganita Sutra-1 'एकाधिकेन पूर्वेण' Ekadhika Purvena / one more than before, the subsequent numerals that is from numeral 'two' to numeral 'nine' to be constructed as arithmetic entities.

As further mathematics is to be work with the help of these numerals, as such appropriate symbols are to be coined presented accepted symbols for these numerals, namely '1, 2, 3, 4, 5, 6, 7, 8, 9' deserve to be adopted as these are having international acceptance by this time and much of our mathematics literature stands composed availing these symbols.

In addition to these nine numerals symbols, the additional symbol '0', designated and known as 'zero', as well accepted for the place value systems and the same as such would get added to the numerals range which would accordingly get extended as (1, 2, 3, 4, 5, 6, 7, 8, 9, 0). However, the values sequence arrangement would make it as a ten steps range beginning with 'zero' and reaching up till '9', being (0, 1, 2, 3, 4, 5, 6, 7, 8, 9).

Here itself it would be relevant to note as that the Ganita Sutra-1 (एकाधिकेन पूर्वेण) answers the inherent question in it as that 'one is more than what?', as that, 'one is more than its previous'. So 0 + 1 = 1, that way also settles the values sequence to start with zero and at next step is to be '1', and there after as per the rule it is to take us to two, and so on, and as such to be the sequential range (0, 1, 2, 3, 4, 5, 6, 7, 8, 9).

The construction of numerals range with the rule of Ganita Sutra-1 as such to be taken up as an exercise to be taken up as an assignment to be completed with all earnestness. The availability of the ten steps long universe range, the arithmetic is to be taken up for its further steps starting with expressions for the numbers in ten place value system.

However, before that a fresh look may be had at the working rule of 'one more than before' for its applied features. Prominent of it being the 'addition operation'. 'To add' means 'to have more'. This, this way, makes the rule 'one more than before', as a principle of 'addition operation'/rule of adding '1' with the previous numeral/number.

## 1. Symbols of 1 and Addition

Here one may have a pause and have a fresh look at the symbols of '1' and 'addition'. It would bring to focus that '1' is of the format of a vertical line. The symbol of addition (+) is availing a pair of lines of the formats of a horizontal line and of a vertical line. As such this symbol of addition (+), avails a pair of lines. The symbol of 1, being of a vertical line so 'one more than before', here is parallel to 'vertical line' getting added to horizontal line' and as such making addition (+). Here, in the context, as such symbolic expression for the rule 'one more than previous' would come to be as that '+' is '1' more than the previous '–'.

## 2. Symbols of Minus

With it 'minus' / '–' is of previous position than that of addition '+'. Though this aspect learning is to be relegated to the subsequent phase and stage of Ganita Sutra-14, under the organization feature, yet, to feed the curiosity, as that 'minus' as of reversal orientation than that of addition along the format of a line can be taken up simultaneously as well, as is the rule of Ganita Sutra-7. Also, it is the symmetry and proportion coordination of negative and positive orientation of a line, as per which under the rule of Ganita Upsutra-1 'आनुरूपेन' Anurupyena 'following the form as it is framed', the minus operation as well would be possible to chase as going from '1' to '0' as being the reverse orientation of reaching from zero to one.

## 3. Symbols of Addition and Multiplication

Now coming back again to the features of the working rule of Ganita Sutra-1 'one more than before', it would clearly emerge as that this is going to be the rule of repeated additions viz. $0 + 1 = 1$, $1 + 1 = 2$, $2 + 1 = 3$ and so on. This feature as such shall be the feature of multiplication operation as that $2 = 1 + 1 = 2 \times 1$ and $3 = 1 + 1 + 1 = 3 \times 1$ and so on. Here itself one may have a pause and have a look at symbol of addition (+) and of multiplication (×). It would come to pointed attention as that both these symbols are availing the pair of lines. However in case of addition '+' the lines are of vertically and horizontal formats while in case of multiplication '×' the lines are diagonally placed. This as such would mean that addition and multiplication are interconnected. The connection is of having are angular positing replacements. It would be like transiting from the sides of a sqsuare to that of the diagonals of a square.

## 4. Symbols of Minus and Division

Further as addition and minus operations are connected so would get connected the multiplication and division operations, as much as that multiplication being the repeated addition and division to be the repeated subtraction. The together look at the symbols of minus (–) and division (/) would also comprehend and to have insight about these symbols for these operations, as much as that the division operation segregates two points (pair of points) as two individual members of the pair.

With it along with the availability of ten numerals symbols there shall be another four operations symbols, namely (–, +, x, /) and the arithmetic basics are to be approached on first principles of Ganita Sutras with the help of these fourteen symbols, namely '0, 1, 2, 3, 4, 5, 6, 7, 8, 9, –, +, ×, /'.

❑ ❑ ❑

# 16

# ARITHMETIC OPERATIONS

## (*i*) Outline Steps

### *1. Addition*

Basic rule:Sutra-1 'One more than before'

### *2. Subtraction*

Basic rule : Sutra-14 'one less than before'

This is a reverse counting rule. It is at the base of the reverse counting principle 'zero is one less than one, one is one less than two, two is one less than three, three is one less than four' and so on.

This becomes the conceptual base of subtraction operation. With its help, the applications of subtraction operation are extended.

On first principle when 4 – 3 is to be worked out, both four and three are chased for their counting positions and value as per the rule of Sutra-14 which shall be taking us 4 = 1 + 1 + 1 + 1 and 3 = 1 + 1 + 1 and accordingly on the number line 4 – 3 = (1 + 1 + 1 + 1) – (1 + 1 + 1) = 1 i.e. single count.

This is in the situation of 3 – 4 shall be leading be 3 – 4 = (1 + 1 + 1) – (1 + 1 + 1 + 1) = –1 i.e. single reverse count.

### *3. Multiplication*

(*i*) As repeated addition (at base)

(*ii*) As repeated addition (at index)

A. By the rule of Nikhilam Sutra

By deficiency from the base

By deficiency from sub base

(*i*) Multiple of base

(*ii*) Part of base (sub base)

B. By Udhrav Sutra

(*i*) Digit wise

(*a*) Single digit multiplication

(*b*) Double digit multiplication

(*c*) Triple digit multiplication

(*d*) Higher digit multiplication

(*ii*) Geometric format

(*a*) For the beginning and last digits

(*b*) For the middle part

(*c*) Multiplication of numbers of value 1 for all digits

(*i*) $1 \times 1 = 1$

(*ii*) $11 \times 11 = 121$

(*iii*) $111 \times 111 = 12321$

(*iv*) $1111 \times 1111 = 1234321$

And so on.....

## (*ii*) Multiplication as Repeated Addition (At Base)

(*a*) By SUTRA-1 'ONE MORE THAN BEFORE'

(*b*) By UPSUTRA-1 'SYMMETRY / PROPORTIONATELY'

4. The value $3 \times 4$ is of three times four. It means four to be added three times i.e. $4 + 4 + 4$.
5. With the help of Rule-1 as $4 = 1 + 1 + 1 + 1$, as such $4 + 4 + 4$ would be

   $(1 + 1 + 1 + 1) + (1 + 1 + 1 + 1) + (1 + 1 + 1 + 1) = 1 + 1 + 1 + 1 + 1 + 1 + 1 + 1 + 1 + 1 + 1 + 1$, i.e. value of 12 counts.

Therefore $3 \times 4 = 4 + 4 + 4 = 12$.

6. The other way to reach at $3 \times 4$ as $4 \times 3$ shall be taking us to the addition of four times three. It would, as per rule-1, make this sum of $3 + 3 + 3 + 3 = (1 + 1 + 1) + (1 + 1 + 1) + (1 + 1 + 1) + (1 + 1 + 1)$ = twelve counts.

## (*iii*) Multiplication as Addition at the Index

(*a*) By SUTRA-1 'ONE MORE THAN BEFORE'

(*b*) By UPSUTRA-1 'SYMMETRY / PROPORTIONATELY'

**7. Illustration 1.**

To multiply $3^4$ by $3^7$

$3 \times 3^7 = 3^{4+7} = 3^{11}$

**8. Illustration 2.**

To multiply $10^9 \times 2^3$

$10^9 \times 2^3 = 2^9 \times 5^9 \times 2^3 \times 5^0 = 2^{9+3} \times 5^{9+0}$

## (*iv*) Multiplication by Rule of Sutra-2 (Case:Base 10)

9. Sutra-2 'All from nine and last from ten'

**Illustration 1.**

Base 10 Illustration $9 \times 7$

Step 1 Deficiency of 9 from 10 is $1 = (-1)$

Deficiency of 7 from 10 is $3 = (-3)$

Put the values in the following format.

| *Number* | *Number value* | *Deficiency from base* |
|---|---|---|
| First number | | |
| Second number | | |
| Product | Cross wise sum | Column wise multiplication |

**Product value**

**It would be**

| *Number* | *Number value* | *Deficiency from base* |
|---|---|---|
| First number | 9 | –1 |
| Second number | 7 | –3 |
| Product | Crosswise Sum | Columnwise multiplication |
| | 7 – 1 = 6 | –1× –3 = 3 |
| Product value | Next place value 6 | Unit place value 3 |
| | 9 × 7 = 63 | |

**10. Illustration 2.**

19 × 18 to be worked out along above format availing 10 as base as

| *Number* | *Number value* | *Deficiency from base* |
|---|---|---|
| First number | 19 | + 9 |
| Second number | 18 | + 8 |
| Product | Cross wise sum | Column wise Multiplication |
| | 18 + 9 = 27 | 9 × 8 = 72 |
| Product value | 27 + 7 = 34 | 2 |
| | 342 | |

## (*v*) Multiplication By Rule Of Sutra-2 (Case : Base 20)

11. Sutra-2 'All from nine and last from ten'

Base 20 = 2 × 10, as such the value which would be reached at next two unit place value is to be multiplied by 2.

Illustration 19 × 18, base 20

| *Number* | *Number value* | *Deficiency from base* |
|---|---|---|
| First number | 19 | –1 |
| Second number | 18 | –2 |
| Product | Crosswise sum | Columnwise multiplication |
| | 18 – 1 = 17 | –1 × –2 = 2 |
| Product value | 17 × 2 = 34 | 2 |
| | 342 | |

## (*vi*) Multiplication by Rule of Sutra-2 (Case:Base 30)

12. Sutra-2 'All from nine and last from ten'

Base 30 = 3 × 10, as such the value which would be reached at next two unit place value is to be multiplied by 3.

**Illustrations** 29 × 28, base 30

| *Number* | *Number value* | *Deficiency from base* |
|---|---|---|
| First Number | 29 | –1 |
| Second Number | 28 | –2 |
| Product | Cross wise sum | Column wise Multiplication |
| | 28 – 1 = 27 | –1 × –2 = 2 |
| Product value | 27 × 3 = 81 | 2 |
| | 812 | |

## (*vii*) Multiplication by Rule of Sutra-2 (Case:Base 40)

13. Sutra-2 'All from nine and last from ten'

Base 40 = 4 × 10, as such the value which would be reached at next two unit place value is to be multiplied by 4.

Illustration 37 × 26, base 40

| *Number* | *Number value* | *Deficiency from base* |
|---|---|---|
| First Number | 37 | –3 |
| Second Number | 26 | –14 |
| Product | Crosswise sum | Columnwise Multiplication |
| | 26 – 3 = 23 | –3 × –14 = 42 |
| Product value | 23 × 4 = 92 | 42 |
| *carry forward | 92+4*=96 | 2 |
| | 37 × 26 = 962 | |

## (*viii*) Multiplication by Rule of Sutra-2 (Case:Base 50)

Sutra-2 'All from nine and last from ten'

Base 50 = 5 × 10, as such the value which would be reached at next two unit place value is to be multiplied by 5.

**14. Illustration.** 50 × 13, base 50

| *Number* | *Number value* | *Deficiency from base* |
|---|---|---|
| First number | 50 | 0 |
| Second number | 13 | –37 |
| Product | Crosswise sum | Columnwise Multiplication |
| | 26 – 3 = 23 | 0 × –37 = 0 |
| Product value | 13 + 0 = 13 | 0 |
| *Base multiple | 13 × 5* = 65 | 0 |
| | 50 × 13 = 650 | |

## (*ix*) Multiplication by Rule of Sutra-2 (Case:Base 100)

15. Sutra-2 'All from nine and last from ten'

Base 100 = 10 × 10, as such the value which would be reached at next two unit place value is to be multiplied by 10.

**Illustration.** 121 × 84, base 100

| *Number* | *Number value* | *Deficiency from base* |
|---|---|---|
| First number | 121 | + 21 |
| Second number | 84 | –16 |
| Product | Crosswise sum | Columnwise multiplication |
| | 84 + 21 = 105 | 21 × –16 = –336 |
| Product value | 105 | –336 |
| | 105 – 4 = 101 | –336 = – 400 + 64 |
| | 10164 | |

❑ ❑ ❑

# 17

# WRITING TABLES 100 × 100

## 1. Counting Line

1. Counting line runs through the counts, points stitched together by the rule 'one more than before' and simultaneously by the rule 'one less than before'.
2. Counting like that as counts points of a counting line goes from one point to the next consecutive point. There is no jump in the process over any of the count point. However, if in the counting process there happens to be a jump over the count points of the counting line then such counting becomes the counting by jumps.
3. Illustratively the counts (1, 3, 5, 7, 9, ——) gives a jump over the counts points (2, 4, 6, 8, ——). It as such is a counting by jump over one count at every step as here would be a counting which shall be taking from '1' to '3' and thereby there would be a jump over count point '2'. And likewise there would be jump at the subsequent steps.

## 2. Rule of Symmetry

4. The counting as counts (1, 2, 3, 4, —) and counting as counts (1, 3, 5, 7, 9, —) are to different counting processes. These are regulated by the rule of symmetry of the Upsutra (Sub Sutra) Anurupyena (आनुरूप्येन)/proportionately/symmetrically.

5. The counting rule of Sutra Ekadhikena Purvena / one more than before, together with the rule of Upsutra Anurupyena / rule of symmetry here shall be making the counting rule as 'one more than before'.
6. The working with the rule (two more than before) shall be taking us from 1 to 1 + 2 = 3 and ahead from '3 to 3 + 2 = 5 and so on'.

## 3. Counting by Jump over Two Consecutive Counts

7. Parallel to the rule of 'two more than before' for counting with jump over one count, there would be a counting rule of 'three more than before' for counting with jump over two counts.
8. As per this rule, the counting steps would be taking from 1 to 1 + 3 = 4, 4 + 3 = 7, 7 + 3 = 10 and so on.
9. Likewise more bigger jumps counting can be symmetrically worked out in terms of Sutra 'Ekadhikena Purvena and Upsutra Anurupyena'.
10. Likewise reverse counting with jumps as well can be worked out as blissful exercises in terms of the chase Ekanuena Purvena Sutra together with Upsutra Anurupyena.

## 4. Table of 1

11. The counts (1, 2, 3, 4, 5, 6, 7, 8, 9, —), by definition, can be accepted as 'table 1' being reached as by Sutra rule 'one more than before'.

## 5. Table of '2'

12. The counts (2, 4, 6, 8, 10, ——)' by definition can be accepted as table 2 being reached as by the Sutra rule 'one more than before together with upsutra rule symmetrical as a working rule 'two more than before'.

## 6. Writing Tables as Counting by Jumps.

13. It would be a blissful exercise to write tables '2, 3, 4, 5, 6, 7, 8, 9, 10' as counting by jumps over 1, 2, 3, 4, 5, 6, 7, 8, 9 jumps respectively and reach at:

Table 2 (2, 4, 6, 8, 10, 12, 14, 16, 18, 20, —)
Table 3 (3, 6, 9, 12, 15, 18, 21, 24, 27, 30, —)
Table 4 (4, 8, 12, 16, 20, 24, 28, 32, 36, 40, —)
Table 5 (5, 10, 15, 20, 25, 30, 35, 40, 45, 50, —)
Table 6 (6, 12, 18, 24, 30, 36, 42, 48, 54, 60, —)
Table 7 (7, 14, 21, 28, 35, 42, 49, 56, 63, 70, —)
Table 8 (8, 16, 24, 32, 40, 48, 56, 64, 72, 80, —)
Table 9 (9, 18, 27, 36, 45, 54, 63, 72, 81, 90, —)
Table 10 (10, 20, 30, 40, 50, 60, 70, 80, 90, 100, —)

## 7. Writing Tables Steps 100 × 100

14. Multiplication operation, as repeated additions, by vedic systems is of very simple mathematical steps. It presumes only the prior learning and remembering of the steps of tables up till 5 × 5, that is of tables of 1 × 2 × 3 × 4 and 5 uptill 5 steps.

15. The bigger digits namely, 6, 7, 8 and 9 are replaced by their deficiencies from base 10, and that is by $\bar{4}$, $\bar{3}$, $\bar{2}$, $\bar{1}$ respectively. With it the numerals range (0, 1, 2, 3, 4, 5, 6, 7, 8, 9) gets replaced by the Vinculum range ($\bar{4}$, $\bar{3}$, $\bar{2}$, $\bar{1}$, 0, 1, 2, 3, 4, 5) for arithmetic operations.

16. With the help of Vinculum range of numerals ($\bar{4}$, $\bar{3}$, $\bar{2}$, $\bar{1}$, 0, 1, 2, 3, 4, 5), for the tables up till 100 × 100, the numbers 1 to 100 are re expressed as follows:

| 01 | 02 | 03 | 04 | 05 | $1\bar{4}$ | $1\bar{3}$ | $1\bar{2}$ | $1\bar{1}$ | 10 |
|---|---|---|---|---|---|---|---|---|---|
| 11 | 12 | 13 | 14 | 15 | $2\bar{4}$ | $2\bar{3}$ | $2\bar{2}$ | $2\bar{1}$ | 20 |
| 21 | 22 | 23 | 24 | 25 | $3\bar{4}$ | $3\bar{3}$ | $3\bar{2}$ | $3\bar{1}$ | 30 |
| 31 | 32 | 33 | 34 | 35 | $4\bar{4}$ | $4\bar{3}$ | $4\bar{2}$ | $4\bar{1}$ | 40 |
| 41 | 42 | 43 | 44 | 45 | $5\bar{4}$ | $5\bar{3}$ | $5\bar{2}$ | $5\bar{1}$ | 50 |

| 51 | 52 | 53 | 54 | 55 | $1\bar{4}\bar{4}$ | $1\bar{4}\bar{3}$ | $1\bar{4}\bar{2}$ | $1\bar{4}\bar{1}$ | $1\bar{4}0$ |
|---|---|---|---|---|---|---|---|---|---|
| $1\bar{4}1$ | $1\bar{4}2$ | $1\bar{4}3$ | $1\bar{4}4$ | $1\bar{4}5$ | $1\bar{3}\bar{4}$ | $1\bar{3}\bar{3}$ | $1\bar{3}\bar{2}$ | $1\bar{3}\bar{1}$ | $1\bar{3}0$ |
| $1\bar{3}1$ | $1\bar{3}2$ | $1\bar{3}3$ | $1\bar{3}4$ | $1\bar{3}5$ | $1\bar{2}\bar{4}$ | $1\bar{2}\bar{3}$ | $1\bar{2}\bar{2}$ | $1\bar{2}\bar{1}$ | $1\bar{2}0$ |
| $1\bar{2}1$ | $1\bar{2}2$ | $1\bar{2}3$ | $1\bar{2}4$ | $1\bar{2}5$ | $1\bar{1}\bar{4}$ | $1\bar{1}\bar{3}$ | $1\bar{1}\bar{2}$ | $1\bar{1}\bar{1}$ | $1\bar{1}0$ |
| $1\bar{1}1$ | $1\bar{1}2$ | $1\bar{1}3$ | $1\bar{1}4$ | $1\bar{1}5$ | $10\bar{4}$ | $10\bar{3}$ | $10\bar{2}$ | $10\bar{1}$ | 100 |

17. The expression of a ten place value expressions for numbers (as of numerals range 0, 1, 2, 3, 4, 5, 6, 7, 8, 9). In terms of the numerals of Vinculum numerals range ( $\bar{4}$, $\bar{3}$, $\bar{2}$, $\bar{1}$, 0, 1, 2, 3, 4, 5) as stand tabulated above, infact lead to the tables writing rule.
18. The table of 99, as such can be written in terms of its corresponding Vinculum value ($10\bar{1}$). As this is a three digits expression, as such the writing of table 99 shall be of three parts of progression in three columns. The progression rule for the first column would be of repeated application of $\bar{1}$, the progression rule of the second column shall be of repeated steps of values '0' ahead, the progression rule for third column digits of table 99 shall be as per the repeated value '1'. The table 99 as such is to be written step by step as follows

| ***Progression*** | | ***1*** | ***0*** | $\bar{1}$ | |
|---|---|---|---|---|---|
| | | ***Third column*** | ***Second column*** | ***First column*** | ***Table Steps value*** |
| Starting step | 1 × 99 | 0 | 9 | 9 | 099 |
| Second Step | 2 × 99 | 0 + 11 | 9 + 09 | 9+ $\bar{1}$ =9–18 | 198 |
| Third Step | 3 × 99 | 1 + 12 | 9 + 09 | 8 – 1 = 77 | 297 |

Like that further steps would follow. Only at the steps where the need would be for carry forward or carry backward that adjustment is to be made.

19. Table of 88 = $1\bar{1}\bar{2}$ shall be having the progression rule for three columns respectively as of value $\bar{2}$, $\bar{1}$ and 1. accordingly the values for the different steps of the tables would sequentially unfold as under.

| *Progression* | | *1* | $\bar{1}$ | $\bar{2}$ | |
|---|---|---|---|---|---|
| | | *Third column* | *Second column* | *First column* | *Table Steps value* |
| Starting step | 1 × 88 | 0 | 8 | 8 | 088 |
| Second Step | 2 × 88 | 0 + 11 | 8 + $\bar{1}$7 | 8 + $\bar{2}$6 | 176 |
| Third Step | 3 × 88 | 1 + 12 | 7 – 16 | 6 – 24 | 264 |

Like that further steps would follow. Only at the steps where the need would be for carry forward or carry backward that adjustment is to be made.

20. Table of 77 = $1\bar{2}\bar{3}$ shall be having the progression rule for three columns respectively as of value $\bar{3}$, $\bar{2}$ and 1. accordingly the values for the different steps of the tables would sequentially unfold as under.

| *Progression* | | *1* | $\bar{2}$ | $\bar{3}$ | |
|---|---|---|---|---|---|
| | | *Third column* | *Second column* | *First column* | *Table Steps value* |
| Starting step | 1 × 77 | 0 | 7 | 7 | 077 |
| Second Step | 2 × 77 | 0 + 11 | 7 + $\bar{2}$5 | 7 + $\bar{3}$4 | 154 |
| Third Step | 3 × 77 | 1 + 12 | 5 – 23 | 4 – 31 | 231 |

Like that further steps would follow. Only at the steps where the need would be for carry forward or carry backward that adjustment is to be made.

21. Like that one can take up any number from 1 to 100 and sequentially tabulate the values of the steps of the table of the chosen number.
22. The interested students may first tabulate the number 1 to 1000 in the vinculum form and then in terms of the vinculum values can write the values of numbers 1 to 1000 up till thousand steps and even more with much ease and in a mechanical manner of the column wise repeated values.
23. The tables even can be extended up till 10000 and even beyond.
24. Division operation

    (*a*) By repeated subtraction:

    (*i*) At base

    (*ii*) At index

    (*b*) By Paravartya Sutra and argumentation.

    (*c*) Straight division of placement values for algebraic expressions formats for single variable.

## A. Division

(*i*) As repeated subtraction (at base).

(*ii*) As repeated subtraction (at index).

❑ ❑ ❑

# 18

# DIVISION AS REPEATED SUBTRACTION

Sutra 14: 'one less than before'

Upsutra-1 'Symmetry/Proportionately'

1. Illustration Dividend = 16, Divisor = 4,

   To find quotient and remainder

   First Step $16 - 4 = 12$

   Second step $12 - 4 = 8$

   Third step $8 - 4 = 4$

   Fourth step $4 - 4 = 0$

As such subtraction four times exhaust the dividend. Therefore quotient = 4 and remainder = 0

2. Illustration 2.

   Dividend = 17, Divisor = 3,

   To find quotient and remainder

   First Step $17 - 3 = 14$

   Second step $14 - 3 = 11$

   Third step $11 - 3 = 8$

   Fourth step $8 - 3 = 5$

   Fifth step $5 - 3 = 2$, which is less than 3.

As such quotient is 5 (equal to repeated steps of subtraction of the divisor from the dividend) And remainder is 2.

❑ ❑ ❑

# 19

# DIVISION BY SUBTRACTION AT THE INDEX

Sutra 14: 'one less than before'

Upsutra-1 'Symmetry/Proportionately'

1. Illustration

   $10^9 / 5^9$

   $10^9 \times 5^9 = (2^9 \times 5^9) / (2^0 \times 5^9)$

   $= (2^{9-0}) \times 5^{9-9})$

   $= (2^9 \times 5^0)$

   $= 2^9$

2. Squaring and cubing
   (*a*) Squaring as multiplication by Urdhrav Sutra
   (*b*) Cubing as repeated multiplication by Urdharav Sutra
   (*c*) Argumentative Straight squaring
   (*d*) Argumentative Straight cubing
   (*e*) Tests of perfect squares
   (*f*) Tests for perfect cubes
3. Square roots and cube roots
   (*a*) By properties of last digits of perfect squares
   (*b*) By properties of the last two digits of perfect squares
   (*c*) By properties of the last three digits of perfect squares

(*d*) By properties of last digits of perfect cubes

(*i*) By properties of the last two digits of perfect cubes

(*ii*) By properties of the last three digits of perfect cubes

4. Decimal expressions for fractions

(*a*) Expressions of finite places of decimal representation

(*b*) Infinite range expressions of decimal representations

(*i*) Recurring ranges

(*ii*) non recurring ranges

5. Fractions

Values of fractions of the time 1/ n, n as a natural number

1/1 = 1

1/2 = .5

1/3 = .33——

1/4 = 1/2 × 2

1/5 = .2

1/6 = 1/2 × 3

1/7 = .(142857) ( 142857)——

1/8 = 1/2 × 4

1/9 = 1/3 × 3

1/10 = 1/2 × 5

1/11 = .(01) (01)——

1/12 = 1/2 × 6

And so on

6. Conversion of last digit of odd (other than five) denominator of the fraction as 9:

(*i*) 9 × 1 = 9, as such if last digit is one, the numerator and denominator can be multiplied by 9.

(*ii*) 3 × 3 = 9, as such if last digit is three, the numerator and denominator can be multiplied by 3.

(*iii*) $7 \times 7 = 49$, as such if last digit is seven, the numerator and denominator can be multiplied by 7.

7. Oscillators

For fractions of recurring decimal expressions, definite oscillators corresponding to recurring features of the decimal expressions can be found by the features of last digit of the denominators.

8. Auxiliary factors

Because of the osculating features of numbers, corresponding auxiliary fractions can be determinant which will help reach at the values of the decimal expression by the process of straight multiplication and division steps in terms of the determinant auxiliary factors.

9. HCF, LCM and Solutions of simple quadratic and cubic equations.

A. By the process of decomposition of the middle terms of products.

The middle-term of the products as a structure permitting split as per the contributions of the crosswise multiplication and summation as per the Udharav Sutra format. This as such, in its reverse steps from the obtained product back to the Udharav steps shall be argumentatively leading to the factors in terms of which HCF, LCM and factorization of algebraic expressions can be reached at.

10. Transition from linear to angular approach.

**A. Sutra:** 2 rule 'all from nine and last from ten' shall be providing the steps for a shift from the linear to angular approach to reach from an line to surface by reaching from a line to an angle, measure in terms of the format for total angle at a point parallel to the circumference of the circle with the given point as its centre.

11. It would be parallel to the place value approach of numbers for a numbers line. The ten place value system

as such shall be providing a measure of 360 degree for the total angle of the surface at a point of the surface. A change to another place value system, say to six place value system, as such shall be leading to '320' degree value measure for the total angle at a point approached by the six place value system. This 320 value measure of six place value system, as such would be equal to $3 \times 6 \times 6 + 2 \times 6 + 0 = 120$ value of ten place value system. Like that different place value system shall be working out different values measures for corresponding transitions from linear to angular set ups.

❑ ❑ ❑

# 20

# NUMBER SYSTEMS

## 1. Counting Number

1. The numbers '1, 2, 3, 4, 5, 6, 7, 8, 9, 10, 11, 12, ——' are designated as counting numbers. The individual numbers are known as counts; count 1, count 2, count 3 and so on.

## 2. Natural Number

2. These counting numbers '1, 2, 3, 4, 5, 6, 7, 8, 9, 10, 11, 12, ——' are also known as natural numbers. Individually, number 1, is known as natural number 1, number 2 is known as natural number 2, number 3 is known as natural number 3 and so on.

## 3. Whole Number

3. The numbers '0, 1, 2, 3, 4, 5, 6, 7, 8, 9, 10, 11, 12, ——' are designated as whole numbers. Individually the number zero is known as whole number zero. The number one is also known as whole number 1 and like that the counting numbers/natural numbers (2, 3, 4, 5, 6, 7, 8, 9, 10, 11, 12, ——) are also individually and collectively are known as whole numbers.

## 4. Numerals

4. The numbers (1, 2, 3, 4, 5, 6, 7, 8, 9) are known as numerals of ten place value system. The number '0' is the place value number of the ten place value number system.

❑ ❑ ❑

# 21

# TEN PLACE VALUE SYSTEM

1. The ten place value system is the arrangement for the values of whole numbers in terms of nine numerals (1, 2, 3, 4, 5, 6, 7, 8, 9) and the placement value (0).
2. Under this arrangement system, ten numbers in this sequence are grouped as a block of numbers. The first block of ten numbers consists of the whole numbers (0, 1, 2, 3, 4, 5, 6, 7, 8, 9). The second block of ten numbers in the sequence is (10, 11, 12, 13, 14, 15, 16, 17, 18, 19).
3. Here it may be noted as that the first block of ten numbers are single digit numbers (0, 1, 2, 3, 4, 5, 6, 7, 8, 9). It may be taken by definition that whole numbers (0, 1, 2, 3) individually are digit (s).
4. The above second block of ten numbers, namely (10, 11, 12, 13, 14, 15, 16, 17, 18, 19) are double digit numbers as that each of these numbers avail two digits for their expression.
5. The placement expression of the digits is as per the values of the places at which the digits are placed. The following pictorial representation is about the way digits are placed and the place values of those digits in ten place value system.

| | | | | |
|---|---|---|---|---|
| — | Fourth place | Third place | Second place | First place |
| — | Value$10^3$ = 1000 | Value $10^2$ = 100 | Value$10^1$ = 10 | Value $10^0$ = 1 |

6. Illustration The single digit numbers (0, 1, 2, 3, 4, 5, 6, 7, 8, 9) shall be having placement only up till the first place.
7. The double digit numbers (10 to 99) shall be having placement covering first place and second place only. The digit in the first place shall be having value of multiple one. While the digit in the second place shall be having value of multiple 10.
8. **Exercise:** The placement expression for the number eleven shall be of two digits as follows:

| — | Fourth place | Third place | Second place | First place |
|---|---|---|---|---|
| — | Value$10^3$ = 1000 | Value $10^2$ = 100 | Value$10^1$ = 10 | Value $10^0$ = 1 |
| | — | — | 1 | 1 |

Eleven = ten plus one, as such 11 = 10 × 1 + 1 × 1 shall be having placement of digit '1' and first place and further also digit '1' at second place as well.

Twenty two = twenty plus two = 20 + 2 = 2 × 10 + 2 × 1, as such the expression

| — | Fourth place | Third place | Second place | First place |
|---|---|---|---|---|
| — | Value$10^3$ = 1000 | Value $10^2$ = 100 | Value$10^1$ = 10 | Value $10^0$ = 1 |
| | — | — | 2 | 2 |

## 1. Integers

9. Corresponding to whole numbers '1, 2, 3, 4, 5, 6, 7, 8, 9, 10, 11, 12, —', there shall be negative whole numbers (–1, –2, –3,–4, –5, –6, –7, –8, –9, –10, –11, –12, —).

   The set of positive whole numbers as well as non zero negative whole numbers (— –4, –3, –2, –1, 0, 1, 2, 3, 4, — ) are designated as integers.

## 2. Rational Numbers

10. The numbers of the form m/n where m and n are integers but n is not zero are designated as rational numbers. As such 1/1 is a rational number while 1 is a counting

number, as well as natural number, and also the whole number and further as well being the integer. Likewise would be the features of 2/1 as comparison to '2'.

## 3. Decimal Numbers

11. The rational number m/10 in its expression of its dividend and remainder separated by 'decimal dot (.)' is designated as decimal expression for the rational number. It also is designated and is known as decimal number.

12. Illustratively the fraction/rational number 13/10 = 10 + 3 = 1 × 10 + 3 =one dividend and three remainder = 1.3 is its decimal expression and as such is designated as the decimal number 1.3.

## 4. Rounding of Numbers

13. When the remainder of m/n is equal to or greater than n/2, value of the dividend is increased by '1'. Otherwise the remainder is left out. This process is the process of rounding of the number 'm/n'.

14. Illustratively 17/3 = 5 × 3 + 2. Here remainder is greater than 3/2. As such as a rounding of process value '1' is added to the dividend '5' which makes it 5 + 1 = 6. As such the rounding of number value 17/3 comes to be '6'.

## 5. Arithmetic Operations

15. Arithmetic operations: Addition, Subtraction, Multiplication, Division.

## 6. Addition Operation

16. Value of counts/counting numbers

    The first count, namely counting number '1' is reached at in one step of 'one more than the previous value '0'', and as such it is taken as of value 'one'.

17. The second count/namely, the counting number '2' is reached at in two steps of the rule 'one more than the

previous one'. The first step takes from '0' to '1' and the second step takes from '1 to 2'.

18. The attainment of the first step which takes from '0 to 1' is the attainment of the addition operation. It is expressed as 0 + 1 = 1. The attainment of the second step which takes from '1 to 2' is the attainment of the addition operation '1+1=2'.
19. This way, the addition operation, sequentially takes from first count to the second count, from second count to third count and so on.
20. It is this feature of the counts which deserves to be comprehended and imbibed fully. Now if we have to add '1' to '2' then, on first principle it would mean that we want to chase and compile, the first step of reaching at from 0 to 1 together with the two steps of reaching from '0 to 2', which shall be amounting to reach at the addition 'Sum' of 1 and 2 as attainment in three steps, which means the count '3'. This being so we shall be writing the addition sum of '1' and '2' as '3'. It is given a mathematical expression of addition of '1' with '2' as that '1+2'=3.
21. It would be a blissful exercise to reach at the addition sum of any of the two numerals on first principles with the help of the working rule of reaching at counts by the principle 'one more than the previous one'.

## 7. Subtraction Operation

22. Parallel to the increasing sequential rule of 'one more than the previous one', would follow the rule of opposite orientation as of decreasing sequence. It would be like going for reverse counting.
23. While the counting/direct counting shall be taking us from '0 to 1' and its reverse counting be taken as taking from '1' to '0'.
24. The attainment of '1' as increasing value step from '0', may be expressed as +1. on the other hand the attainment

of '0' starting with '1'as reverse counting step shall be bringing in the negative value for '1', which may be expressed as '1'.

25. This process of reaching from '1' to '0' shall be taken as the subtraction operation as per which '1' plus '–1' = 0. It in its simplified form is expressed as (1) – (1) = 0 its equivalent is 1 – 1 = 0.

## 8. Multiplication

26. Multiplication may be taken as repeated addition. When it is written 2 × 3 it means 2 to be multiplied by 3, or other way around as that '3 to be multiplied by 2'.
27. The multiplication value of '2 × 3' would mean that '2 to be added three times, i.e. 2 + 2 = 4 and ahead 4 + 2 = 6. It is of the expression 2 + 2 + 2 = 6 = 2 × 3.
28. The other way round it also means 3 multiplied by 2, i.e. 3 to be added twice, i.e. 3 + 3 = 6.
29. Likewise to comprehend the multiplication steps on first principles one may attempt multiplication of two numerals say 4 × 7 or 8 × 9 and so on.

## 9. Division Operation

30. Parallel to addition and subtraction as being of reverse orientations, the multiplication and division as well may be taken as of reverse orientation.
31. As multiplication is repeated addition, the division may be taken as repeated subtraction.
32. The value of 16/4 may be reached at in sequential steps of subtraction of value 4 from the value 16 till the values comes to be 0 or less then 4. the working steps for 16/4, as repeated subtraction would follow as under

| | |
|---|---|
| At first step | 16 – 4 = 12 |
| At second step | 12 – 4 = 8 |
| At third step | 8 – 4 = 4 |

At fourth step $4 - 4 = 0$

As such 16/ 4 attains '0' in four steps, as such the value of the division of 16 by 4 comes to be '4', which is equivalent to the number of steps.

33. The working steps for 19/3 shall be as follows:

| | |
|---|---|
| At first step | $19 - 3 = 16$ |
| At second step | $16 - 3 = 13$ |
| At third step | $13 - 3 = 10$ |
| At fourth step | $10 - 3 = 7$ |
| At fifth step | $7 - 3 = 4$ |
| At Sixth step | $4 - 3 = 1$ which is less than 3 |

As such the division of 19/3 comes to be '6' as complete steps together with '1' as remainder at the step.

## 10. Arithmetic Operations:Upon Integers

34. The concept of absolute value number is to be introduced first before the Arithmetic operations skills are to be introduced.
35. For the concept of absolute value number, the format of a line may be availed. On the scale, absolute value '1' may be taken as one unit. When this unit is on the west to east direction, it is taken as +1 and when this unit is on the east to west direction from the origin, it is taken as –1. Like that absolute value '2' makes +2 and –2 i.e. positive number 2 and negative number 2.

## 11. Step Six To Nine

36. The sixth to ninth letter of Ganita Sutra-1 compose the formulation 'केन'.
37. The formulation 'केन' has expression range of reasoned questions for format chase.
38. This four fold formulation of letters (*i*) 'क्' (*ii*) 'ए' (*iii*) 'न्' and (*iv*) 'अ' is to be taken parallel to four fold manifesta-

tions within creator space (4 space) 'क्'.

39. It shall be sequentially leading to the values (*i*) pair of dimensions of order of 4-space, which shall be leading to 4 + 4 – 2 = 6 / 6-space domain (*ii*) Further 6 + 6 – 4 = 8 is to be of the 8 space domain (*iii*) 8 + 8 – 6 = 10 / Par Braham phase and stage, but as it as such transcends beyond linear digits range, it shall be taking us back to the ten components creative boundary of 5 space, and finally (*iv*) 5 + 5 – 3 = 7 / 7 space shall be attaining unity state of existence phenomenon.
40. One may have a pause here and permit the transcending mind to transcend through the transcendental feature as much as that 7 space is to the play the role of dimension of 9 space.
41. Further as that 16 letters range of the text of Ganita Sutra-1 'एकाधिकेन पूर्वेण', accordingly, accepts a pair of formulations of 9 and 7 letters each.
42. With it, it would be desirable before proceeding further to the next 7 steps i.e. from steps 10 to 16 for the attainment of enlightenment along the formats of Ganita Sutras, the needed shift would be from artifices approach of Sankhiya Nishtha to geometric approach of Yoga Nishtha.
43. With this in mind, hence forth, first of all the over view of the approach to the discipline of the geometry is being taken up, to be followed by the query as 'how to start learning and teaching of the discipline of geometry to young mind'.
44. The geometric bodies range, point, line, angle, triangle, circle, square, sphere, cube, hyper cube 4, pentagon, hyper-cube 5, hexagon, hyper cube 6, 6-space is being taken up for proper insight then the geometric envelope of cube, manifestation of cube in 4 space, transcendence through manifestation and finally the transcendental world are being visited here under.

❑ ❑ ❑

# 22

# OVERVIEW OF THE APPROACH TO THE DISCIPLINE OF THE GEOMETRY

## 1. Introductory

1. Vedic Systems approach is multi-dimensional.
2. Vedas avail real spaces as Organization formats
3. Parallel to Numerals of the place value system emerge corresponding real-dimensional spaces.
4. These real dimensional spaces are of sequential dimensional orders.
5. Our well known three dimensional bodies are of linear dimensional set-ups, that is, here one space plays the rule of dimension.
6. The Geometry which restricts itself uptil three space, that is, uptil solids, is excepted as a linear order Geometry. It is design-ated and is known as "Rekhaganit"/ The Mathematics of line.
7. Vedic Geometry, as such, is not restricted uptil linear dimensional order. Beyond that special dimensional order of four space, solid order of five space, and hyper solid orders from six space onwards as well are the real interest of Vedic Geometry.
8. However as, in the present first volume, the chase is restricted only uptil the linear order as such the focused exposure is entered upon solids, and of these as well, only

upon the lines, surfaces and solids with the running formats of interval, square and cube as well as of Point, line, circle and sphere.

## 2. Artifices of Numbers and Dimensional Frames

9. Vedic Processing systems are of two fold features, designated and known as Sankhya Nishta and Yoga Nishtha.
10. Sankhyanishtha presume the existence of Geometric formats and avails values of numbers.
11. Yoganishtha presumes the existence of numbers and avails the values of the Geometric formats.
12. Of it the specific applied values of numbers and Geometric formats being availed as of the artifices of number and dimen-sional spaces.
13. The artifices of numbers are chased presuming the existence of dimensional spaces, and on the other hand the dimensional frames are chased presuming the existence of artifices of numbers.
14. Standing restricted for the chase only uptil the linear order three space, the dimensional frame made use of three dimensions as three axes of three lines format.
15. Corresponding to it the artifices of first three numbers namely (1, 2, 3) / (1, 1 +1, 1+1+1) / (1,1 × 1,1 × 1 × 1) / ($1^1$, $1^2$, $1^3$) / [ (1), (1, 1), (1, 1, 1)]

## 3. Ganita Sutras Approach

16. While going for arithmetic chase, the Geometric formats are presumed.
17. It would mean that, it is being presumed that Ganita Sutras rules are as per the features of Geometric formats being Sequentially availed from first Sutra to the last Sutra, as well as from first Upsutra till the last Upsutra.
18. With availability of the sequential Geometric formats, the values of arithmetic starting with sequential features of artifices of numbers are worked out.

19. It would mean that arithmetic is being done on the Geometric formats.
20. On the other hand while the Geometry, add in particular Rekhaganit/Mathematics of line is being worked, it is being presumed that values of numbers, and in particular the features of artificies of numbers are available.
21. It is in this background that the working rule of Ganita Sutra 1 "One more than before" shall be sequentially unfolding "points as zero degree bodies, lines as first degree bodies, planes as two degree bodies, solids as four degree bodies, hypersoldis four as four degree bodies and so on."
22. Ganita Upsutra-1 with its rule of Symmetry and Proportion chase by following the forms as these are friend shall be sequentially taking us from lines to parallel lines to angles, to triangles, to quadrilaterals, to pentagons, and so on to circles, to spheres, to hyperspheres and so on.
23. Ganita Sutra-2 with the working rule of "all from Nine and last from ten", making available ten place value system for Organizations of artifices of numbers and for values of numbers, shall be correspondingly leading to the chase of Geometric formats through angles.

Here it may be relevant to mention for pointed attention.

## 4. Total Angle Value in Different Place Values

Parallel to ten place value system, total angle at a point in the surface may be had in terms of different place value systems. On six place value system as there are five numerals as such the value of total angle for all the four quarters shall be 4×(556)=120. However as in six place value system "6" = "10", as such it shall be 4 × (50) = "320" in six place value system. Like that the value may be computed in different place value system.

The sequential order of Ganita Sutras as along artifices 1 to 16, and corresponding to it the parallel Geometric formats thereof, shall be sequentially leading as Ganita Sutra-1 organizing features of line, sequential order of line, linear dimensional order. Ganita

Sutra-2 shall be leading to Surfaces, special orders, angular coverage parallel to different place value system. Ganita Sutra-3 leading to solid order with transition from horizontal plane to vertical plane. Ganita Sutra-4 shall be taking a step ahead to a four space with reflection operation and working with half dimensions and so on. A step ahead would be a transcends from avyakta to aviakto-avyakat, that is from special order for space to solid order five space, where the zero space and zero space content shall be having their roles to play, which symmetrically is to be availed for six space values to which Ganita Sutra-6 leads to. Beyond that, that is from Sutra-7 onwards is to be self referral feature of plus one space and negative one space are two simultaneously come into play and with it the chase is to really enter the real spaces features of hyper dimensional orders, which chase is being left to be taken of in the other volumes.

❑ ❑ ❑

# 23

# HOW TO START LEARNING AND TEACHING OF THE DESCIPLINE OF GEOMETRY TO THE YOUNG MINDS

## SYMBOLS

1. Geometric bodies can well be better approached with the help of symbols.
2. Here below are being tabulated symbols of some of the geometric entities/bodies.

| Sr. No. | Geometric body | Symbol |
|---|---|---|
| 1 | Point |  |
| 2 | Interval |  |
| 3 | Circle | 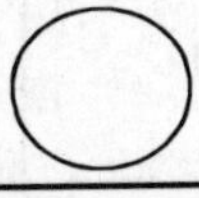 |
| 4 | Square |  |
| 5 | Cube |  |
| 6 | Sphere |  |
| 7 | Four space |  |

| 8 | Five space |  |
|---|---|---|
| 9 | Six space |  |
| 10 | Hyper cube 4 | H |
| 11 | Hyper cube 5 | Q |
| 12 | Hyper cube 6 | b |
| 13 | Triangle |  |
| 14 | Hexagon | 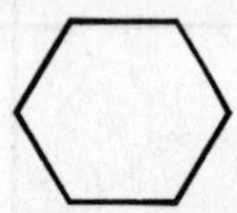 |
| 15 | Cylinder |  |
| 16 | Cone |  |

# 24

# GEOMETRIC BODIES RANGE: POINTS

1. One shall sit comfortably and permit the transcending mind to chase the setup of the point.
2. Point may be chased as a 0-space body.
3. Point can be chased as a sphere.
4. Point can be chased as a sphere line.
5. Point also can be chased as sphere area.
6. Still further point can be chased as a sphere body of 3-space.
7. Likewise point can be chased as a sphere body of 4-space.
8. Likewise point can be chased as a sphere body of 5-space.
9. Likewise point can be chased as a sphere body of 6-space.
10. It would be a blissful exercise to reach the distinct gushing features of points of different dimensional spaces.

❑ ❑ ❑

# 25

# LINE

1. Line is a 1-space content display.
2. Line is the expression of the path of a moving point.
3. It would be a blissful exercise to permit the transcending mind to chase the path of moving point:

   (*i*) With in plain. (*ii*) With in 3-space.

   (*iii*) With in 4-space. (*iv*) With in 5-space.

   (*v*) With in 6-space.
4. Line as interval is a representative regular body of 1-space, and as such line is of the values of 1-space domain.
5. Line as circumference of circle is in the role of boundary of 2-space as here 1-space is in the role of boundary.
6. Line as axis is in the role of linear dimension of 3-space and pair of axes put it in the role of boundary while all the three axis put themselves in the role of domain of 3-space.
7. Line, how so ever long is of 0 area, and volumme.
8. Line it self along manifestation format is of 4-folds with –1 space as its dimension fold.
9. Line along the transcendence range coordinates the transcendence and ascendance through the manifestation layers and there by becomes the spectrum format for Divya Ganga flow.

❑ ❑ ❑

# 26

# ANGLES

1. 'Angle' is space strip between a pair of lines.
2. It is a surface within a pair of boundary lines.
3. It is a 2-space setup within a pair of axes.
4. It is a point, a pair of lines and a surface in between the pair of lines.
5. It is a setup of 0-space, 1-space and 2-space entities.
6. These 3 setups together also constitute a measuring line of 2-space.
7. With 0-space as dimension, 1-space as boundary and 2-space as domain raises and expectation for 3-space as origin.
8. It would be a blissful exercise to chase setup of an angle as a triangle as half rectangle.
9. One shall sit comfortably and permit the transcending mind to chase the setup of a triangle, as a set of 3 angles.
10. Further it would be blissful to chase triangle as a printout of 3-space in 2-space with a pair of facets, a transcendence phenomena taking us from 2-space to its dimension, being 0-space and further to a dimension of dimension of 2-space, that is to (–2 space).

❑ ❑ ❑

# 27

# TRIANGLE

1. One shall sit comfortably and permit the transcending mind to chase the setup of a triangle.
2. It is a setup of 3 corners points, 3 edges and one in between surface area.
3. It would be relevant to note that a cube is a representative regular body of 3-space and 3-space admits 7-geometries of which three are negative geometries, three are positive geometries and one is a neutral geometries.
4. It would be blissful to permit the transcending mind to chase different geometry components of setup of a triangle as printouts of different geometries of 3-space.
5. Triangle itself as a print out of 3-space setup shall be providing an envelope for the center of triangle/Origin of 2-space/origin fold of manifestation layer (0, 1, 2, 3).
6. It would be a blissful exercise to permit the transcending mind to chase triangle as origin fold of the manifestation layer (0, 1, 2, 3).
7. It would further be a blissful exercise to chase the phenomena of triangle being the first setup of lines enclosing a space.
8. It would further be a blissful exercise to chase surface/2-space as dimension of 4-space, being the origin of 3-space.

9. It would further be a blissful exercise to permit the transcending mind to phenomena of spatial order (as surface of triangle) leading to a solid order (In terms of three linear edges together as a three linear axes of 3-space).
10. It would further be a blissful exercise to reach from edges as a linear order to spatial order inside as surface and then reaching at solid order at the boundary.

❑ ❑ ❑

# 27
# CIRCLE

1. Circle is a setup of 2-space.
2. Circle avails 1-space in the role of boundary fold as circumference of the circum.
3. There is permanent bond between the roles of 1-space as diameter as domain fold and as circumference as a boundary fold.
4. It would be a blissful exercise to permit the transcending mind to chase the common value of the above permanent bond between different roles of 1-space within 2-space.
5. As 2-space accepts a 3-space setup for the origin seat, as such the values of the above permanent bond have a sequential increases up-till the availability of the geometric format of cube.
6. It would be a blissful exercise to define and chase hyper-circle as a setup of points at a common distance from the origin.
7. One shall sit comfortably and permit the transcending mind the transcendental phenomena of sequential increase for the values of hyper-circles 1 to 7.
8. It would further be a blissful exercise to reach at transition from increasing feature of values of hyper cubes 1 to 7

Suddenly in to the decreasing feature of values of hyper cubes 8 onwards.

9. One shall sit comfortably and permit the transcending mind to chase the cause for the above sudden change from increasing values to decreasing values.
10. It would be a blissful exercise to chase and comprehend the above cause as a transition beyond the range of 7 geometries of linear order setup of the cube to the spatial order beyond that range coming in to play.

❑ ❑ ❑

# 29
# SQUARE

1. Circle and square are the representative regular bodies of 2-space.
2. Circle is a member of sequence of dimensional bodies: Circle, sphere, Hyper sphere 4, Hyper sphere 5 and so on.
3. Square is member of sequence of dimensional bodies: Interval, Square, Cube, Hyper cube-4, Hyper cube-5 and so on.
4. The domain boundary ration formulation for the sequence (Interval, Square, Cube, Hyper cube-4, Hyper cube-5 and so on) is

$$A^n : 2nA^{n-1}, n = 1, 2, 3, 4, 5- - -.$$

5. It would be a blissful exercise to chase the setups of (Interval, Square, Cube, Hyper cube-4, Hyper cube-5 and so on) in terms of the above formulation.
6. It would further be a blissful exercise to chase and comprehend as that the formulation $A^n : 2nA^{n-1}$, n = 1, 2, 3, 4, 5- - -. Also holds for the sequence (Circle, sphere, Hyper sphere 4, Hyper sphere 5 and so on).
7. It would further be a blissful exercise to simultaneously chase the setups of circle and square.

8. It would be interesting to comprehend as that the boundary of circle is a continuously closed setup while the domain of square is a continuously closed setup.
9. One shall sit comfortably and permit the transcending mind to chase the boundary of square as a synthetic setup of 4-components stitched at 4 corners.
10. It would further be a blissful exercise to chase the way the domain of circle a stitched as 4-components with in a dimensional frame of pair of axes of 3-space.

❑ ❑ ❑

# 30

# SPHERE

1. Cube and sphere are the representative regular body of 3-space.
2. Sphere is a member of the sequence 'Circle, Sphere, Hyper Sphere-4, Hyper Sphere-5 and so on.
3. The domain of the sphere is wrapped within a surface.
4. The enveloping surface is continues boundary.
5. Being enveloped within a spatial boundary of continuous surface sheet, its domain can be split into many-many components.
6. Of these splits, 6-components split and 8-components split are of spatial attention, as on the one end, it is a sequential step ahead of the pervious member of the sequence, and on the other end it needs to the next member of the sequence.
7. It would be a blissful exercise to view this placement position of the sphere in between circle and hyper sphere-4, parallel to the placement of the cube in between the square and hyper cube-4.
8. The sequence 'Square, cube and Hyper cube-4 etc' yields boundary components values as 4, 6, 8 and so on.
9. Parallel to it, the sequence 'Circle, cube, Hyper cube-4 etc' Yields domain components values as 4,6,8 and so on.
10. However, both the sequence still have distinguishing features which deserve to be glimpsed and comprehended fully.

❑ ❑ ❑

# 31
# CUBE

1. The enveloping boundary of cube is stitched in terms of 26 boundary components, namely 8 corner points, 12 edges and 6 surfaces.
2. If the edge of the cube is taken as a component interval of 2 parts, that is as a set of 3 points (2 end points and middle points) and a pair of lines in between, that is, as set of 5-components, them the geometric setups due to first, first two and all the three axes shall be of $5^1$, $5^2$ and $5^3$ components respectively.
3. It would be a blissful exercise to chase the setups due to first, first two and all the three axes of the cube with its edges being of 3, 4, 5 and so on parts.
4. It would further be blissful to reach at a common formulation for the above geometric setups.
5. One shall sit comfortably and chase sequentially the interval of 2, 3, 4, 5 and so on parts and one would comprehend that these shall be of the values and order $NA^1+(N+1)A^0$ for $A = 1$ and $N = 1, 2, 3, 4, 5$ and so on.
6. These values respectively shall be 3, 5, 7, 9, 11 and so on.
7. As such the values and order of the setups of intervals of different parts shall be respectively 3, 5, 7, 9, 11 and so on.

8. It would be relevant to note that these values are parallel to values and orders of the geometries of the dimensional spaces, 1,2,3,4,5 and so on being respectively 3, 5, 7, 9, 11 and so on.
9. It would be blissful to glimpse and chase the geometrics setup of squares of intervals of 1,2,3,4,5 and so on as being $3^2, 5^2, 7^2, 9^2, 11^2$ and so on.
10. It further would be very blissful glimpse and chase the geometrics setup of cubes of intervals of 1,2,3,4,5 and so on as being $3^3, 5^3, 7^3, 9^3, 11^3$ and so on.

❑ ❑ ❑

# 32

# HYPER CUBE-4

One shall sit comfortably and permit the transcending mind to glimpse and chase the geometric setup of Hyper cube-4 and comprehend the following features of hyper cube-4:

1. Hyper cube-4 is a representative regular body of 4-space.
2. Hyper cube-4 is a manifestation layer of 4 folds (2,3,4,5).
3. Two space plays the role of dimension fold and as such the setup is designated as of spatial order.
4. Three space plays the role of boundary fold and as such the setup is designated as of solid boundary.
5. Four space plays the role of domain fold and as such the setup is designated as of Hyper domain.
6. Five space plays the role of Origin fold and as such the setup is designated as of transcendental origin.
7. One shall permit the transcending mind to glimpse and comprehend the above format of Hyper cube-4 as a lotus seat of 8 petals of 4 head lord equipped with a pair of eyes in each head and a transcendental heart.
8. Lord Brahma, the four head lord, is the lord of 4-space.
9. Four space as such is designated as creator's space.
10. Lord Brahma is the Creator the supreme.

❑ ❑ ❑

# 33

# PENTAGON

1. Pentagon is a member of the sequence 'Triangle, Rectangle, Pentagon, Hexagon and so on.
2. One feature of this sequence is the value of the space enclosed being within 3, 4, 5, 6 and so on sides.
3. The values '3,4,5,6 and so on' are parallel to the values and order of '3-space, 4-space, 5-space, 6-space and so on.
4. It would be a blissful exercise to permit the transcending mind to glimpse 'Triangle, Rectangle, Pentagon, Hexagon and so on as print outs of representative regular bodies of '3-space, 4-space, 5-space, 6-space and so on.
5. One shall sit comfortably and permit the transcending mind to chase the setup of pentagon as a printout of representative regular body of 5-space.
6. It would further be blissful to glimpse, comprehend and chase the external angles of pentagon as of the value (5–2) Pie.
7. One shall sit comfortably and permit the transcending mind to glimpse, comprehend and chase the bonds of artifices 5 and 3 as 5-space and 3-space being interconnected as domain fold and dimension fold.

8. Further it would be blissful to glimpse, comprehend the external angles of pentagon being the expression of 3-space in the role of dimension of 5-space.
9. It would be blissful to chase artifice 5, pentagon and 5-space of parallel features.
10. It would be a blissful exercise to internally connect all the corners of the pentagon and to glimpse the formation of internal pentagon, as a transcendental phenomena as that there shall always emerging a pentagon within a pentagon.

❑ ❑ ❑

# 34

# HYPER CUBE-5

1. One shall sit comfortably and chase the features of artifice-5, 5-space, pentagon and Hyper cube-5.
2. Artifice-5 has 5-artifices, 5-space has 5-dimensions, pentagon has 5 sides and Hyper cube-5 is representative regular body of 5-space in 4-space.
3. One shall sit comfortably and permit the transcending mind to comprehend the parallel features of artifice-5 and of 5-space as parallel formats of Sankhiya Nishta and Yoga Nishta processing formats.
4. One shall sit comfortably and further permit the transcending mind to glimpse and comprehend pentagon-5 as print out of Hyper cube-5 in spatial dimension of 4-space.
5. The features of the pentagon that when all its corners are connected with all other corners, they envelop the center of the pentagon with a pentagon, to be designated as inner envelop.
6. One shall sit comfortably and permit the transcending mind to continue in a prolonged state of deep trans and glimpse this feature of inner pentagons envelopes, one within the other, as a never ending sequence.

7. It would be a blissful exercise to comprehend the values of the phenomena of never ending sequence of inner pentagons.
8. It would be a blissful to comprehend the parallel feature of inner sequences and of inner folds of the transcendental worlds.
9. It would further be a blissful exercise to chase the inner folds of Upper part of transcendental worlds along the format of Hyper cube-5.
10. It would further be a blissful exercise to chase the inner folds of Lower part of transcendental worlds along the format of Hyper cube-5.

❑ ❑ ❑

# 35
# HEXAGON

1. Hexagon is a member of the sequence 'Triangle, Rectangle, Pentagon, Hexagon and so on.
2. One feature of this sequence is the value of the space enclosed being within 3, 4, 5, 6 and so on sides.
3. The values '3,4,5,6 and so on' are parallel to the values and order of '3-space, 4-space, 5-space, 6-space and so on.
4. It would be a blissful exercise to permit the transcending mind to glimpse 'Triangle, Rectangle, Pentagon, Hexagon and so on as print outs of representative regular bodies of '3-space, 4-space, 5-space, 6-space and so on.
5. One shall sit comfortably and permit the transcending mind to chase the setup of Hexagon as a printout of representative regular body of 6-space.
6. It would further be blissful to glimpse, comprehend and chase the external angles of Hexagon as of the value (6 – 2) Pie.
7. One shall sit comfortably and permit the transcending mind to glimpse, comprehend and chase the bonds of artifices 6 and 4 as 6-space and 4-space being interconnected as domain fold and dimension fold.

8. Further it would be blissful to glimpse, comprehend the external angles of pentagon being the expression of 4-space in the role of dimension of 6-space.
9. It would be blissful to chase artifice 6, Hexagon and 65-space of parallel features.
10. It would be a blissful exercise to internally connect all the corners of the Hexagon and to glimpse the formation of internal Hexagon, as a transcendental phenomena as that there shall always emerging a Hexagon within a Hexagon.

❑ ❑ ❑

# 36

# HYPER CUBE-6

1. One shall sit comfortably and chase the features of artifice-6, 6-space, Hexagon and Hyper cube-6.
2. Artifice-6 has 6-artifices, 6-space has 6-dimensions, Hexagon has 6 sides and Hyper cube-6 is representative regular body of 6-space in 4-space.
3. One shall sit comfortably and permit the transcending mind to comprehend the parallel features of artifice-6 and of 6-space as parallel formats of Sankhiya Nishta and Yoga Nishta processing formats.
4. One shall sit comfortably and further permit the transcending mind to glimpse and comprehend Hexagon as print out of Hyper cube-6 in spatial dimension of 4-space.
5. The features of the Hexagon that when all its corners are connected with all other corners, they envelop the center of the Hexagon within a Hexagon, to be designated as inner envelop.
6. One shall sit comfortably and permit the transcending mind to continue in a prolonged state of deep trans and glimpse this feature of inner Hexagon envelopes, one within the other, as a never ending sequence.

7. It would be a blissful exercise to comprehend the values of the phenomena of never ending sequence of inner pentagons.
8. It would be a blissful to comprehend the parallel feature of inner sequences and of inner folds of the Self-referral worlds.
9. It would further be a blissful exercise to chase the inner folds of Upper part of Self referral worlds along the format of Hyper cube-6.
10. It would further be a blissful exercise to chase the inner folds of Lower part of Self-referral worlds along the format of Hyper cube-6.

❑ ❑ ❑

# 37

# 6-SPACE

1. 6-space is a self-referral state.
2. Sun is a 6-space body.
3. Human frame accepts a 6-space measuring rod.
4. 6-space measuring rod is constituted by representative regular bodies of Hyper cube-1 to Hyper cube-6.
5. Lord Vishnu is the presiding deity of this measuring rod.
6. Lord Brahma is the presiding deity of the measure of this measuring rod.
7. Along the format of this measuring rod is manifested the Shad Chakras format of Human frame.
8. The 6 × 6 matrix depicts the roles of 6-space/artifice-6 as under:

| | | | | | |
|---|---|---|---|---|---|
| 1 | 2 | 3 | 4 | 5 | 6 |
| 2 | 3 | 4 | 5 | 6 | 7 |
| 3 | 4 | 5 | 6 | 7 | 8 |
| 4 | 5 | 6 | 7 | 8 | 9 |
| 5 | 6 | 7 | 8 | 9 | 10 |
| 6 | 7 | 8 | 9 | 10 | 11 |

9. One shall sit comfortably and chase the formulation:

NVF (ZERO) = 64

10. One shall sit comfortably and further chase the formulation:

NVF (UNIT) = 64

❑ ❑ ❑

# 38

# GEOMETRIC ENVELOPE OF CUBE

1. One shall sit comfortably and permit the transcending mind to chase the way the geometric envelope of cube is stitched.
2. One shall comprehend well that in each of the 8 corner points, there is embedded a three dimensional frame of half dimensions.
3. Also there is embedded also three dimensional frame of full dimensions at the center of the cube.
4. One shall permit the transcending mind the chase of the stitching steps at the corner points of the cube.
5. One shall sit comfortably and permit the transcending mind to chase the phenomena of these stitching steps.
6. One shall also pose to one self as to the face and stage of the phenomena of manifestation of cube prior to the emergence of the stitching steps at the 8 corner points.
7. One shall further pose to one self as to the features of the phenomena of manifestation of the cube prior to the stitching steps in terms of 12 edges of the cube.
8. One shall also pose to one self as to how the edges with features of simultaneous having opposite orientations to

provide half-dimensional frames emanating from the corners points of the cube.

9. Further one shall permit the transcending mind to chase the phenomena of manifestation of the cube prior to the face and stage of six sitting steps in terms of six surface plates of the cube.
10. One shall pose to one self the set of features of the cube being free of its geometric envelope.
11. One shall permit the transcending mind to chase NVF (Cube) = 31 as a Value of 26 components of the envelope of a domain (Volumme) with a three dimentional frame embedded at its center together consvuting 31 geometric components of the geometric setup of the manifested cube.
12. One shall further permit the transcending mind to chase NVF (Cube) = NVF (Cave).

❑ ❑ ❑

# 39

# GEOMETRIES BODY RANGE OF MANIFESTATION OF CUBE IN 4-SPACE

1. The Vedic mathematics teacher shall help be young minds to chase the manifestation of cube within 4-space.
2. Simultaneously the Vedic Mathematics teacher shall also help the way 4-space is released at the center of the cube and manifest as hyper cube-4.
3. Hyper cube-4 is a spatial order setup of 4-dimension within solid boundary of 8 components.
4. Cube admits a cut into 8 sub cubes of half dimensions with center of the cube as common point of all the eight sub cubes.
5. The centre of the cube as such is enveloped by the eight corners points of eight sub cubes having common placements at centre.
6. The centre as a point, 0-space, and its boundary as (–1) space, puts the corners point of eight sub-cubes in such orientation placements along the boundary of 'Centre'/ Point/0-space because of which the three space dimensional frames imbedded in the corners points acquire the corresponding orientation.

7. Because of this, the manifestation features of cube make it a manifested body of 1-space as dimension, 2-space as boundary, 3-space as domain and 4-space as origin.
8. While hyper cube-4 its self is a spatial order space of solid boundary and transcendental origin.
9. It would be a blissful exercise to permit the transcending mind to chase the manifestation of cube within Hyper cube-4 as representative regular body of 4-space.
10. It also would be blissful exercise to permit the transcending mind to chase the released of 4-space at centre of the cube and its emergence and manifestation as Hyper cube-4.

❑ ❑ ❑

# 40

# GEOMETRIES BODY RANGE TRANSCENDENCE THROUGH MANIFESTATIONS

1. The phenomena of manifestation of cube with in hyper cube-4 leads to emergence of cube as a manifestation layer (1,2,3,4) along Hyper cube-4 (2,3,4,5) as base.
2. Simultaneously, the emergence of hyper cube-4 at centre of cube, as such, makes out a phenomena of transcendence through manifestation layer (1,2,3,4).
3. This transcendence phenomena happens because of 5-space/ transcendental world/ Hyper cube-5 as origin fold of hyper cube-4, that is, of manifestation layer (2,3,4,5).
4. The emergence and manifestation of cube with in 4-space is transcendence process with cube it self, as representative regular body of 3-space can play they role of solid dimension of 5-space.
5. One may take manifestation format being of 4-folds as (1,2,3,4), while the transcendence process being of 5-folds (1,2,3,4,5).
6. One may take transcendence process as going from the setup of cube to the setup of hyper cube-4.
7. Further one may take the ascendance process has a process of going from hyper cube-4 to cube capable of

playing the role of dimension of the transcendental worlds.

8. The ascendance process as such takes ascendance from the spatial order to solid order.
9. On the other hand, the transcendence process takes from 3-space domain (body)/(Cube), to 4-space domain (body)/(Hype cube-4).
10. To sum up, the transcendence takes from lower domain to hire domain while ascendance takes from lower dimension order to hire dimension order.

❑ ❑ ❑

# 41

# GEOMETRIES BODY RANGE TRANSCENDENTAL WORLDS

1. Transcendental world is a 5-space setup.
2. It is solid order setup.
3. It is a setup within 5 solid dimensions frame.
4. It has seat at the origin of creator's space (4-space).
5. Hyper cube-5 is a representative regular body of 5-space and as such the transcendental world manifest along the format of Hyper cube-5.
6. Transcendental world emanates from the origin of Creator's space and manifest as hyper cube-5 within creator's space along its manifestation format.
7. The transcendental world manifest as a sky within a space.
8. Lord Brahma is the over load of 4-space (Creator's space), while Lord Shiv is the over load of 5-space (Transcendental world).
9. One shall sit comfortably and permit the transcending mind to chase the phenomena of lord Brahma multiplying as ten brahmas.
10. It further would be a blissful exercise to chase the phenomena of emergence of lord Brahma along each of the ten direction of solid dimension of the transcendental world.

❑ ❑ ❑

# 42
# HYPER CUBES SEQUENCE

| | | |
|---|---|---|
| 1 | $1^2$ | 1-Space/Interval |
| 2 | | First end point |
| 3 | | Second end point |
| 4 | $2^2$ | 2-Space/Square |
| 5 | | First boundary line |
| 6 | | Second boundary line |
| 7 | | Third boundary line |
| 8 | | Fourth boundary line |
| 9 | $3^2$ | 3-Space/Cube |
| 10 | | First Surface plate |
| 11 | | Second Surface plate |
| 12 | | Third Surface plate |
| 13 | | Fourth Surface plate |
| 14 | | Fifth Surface plate |
| 15 | | Sixth Surface plate |
| 16 | $4^2$ | 4-Space/Hyper Cube 4 |
| 17 | | First Solid boundary component |

| | | |
|---|---|---|
| 18 | | Second Solid boundary component |
| 19 | | Third Solid boundary component |
| 20 | | Fourth Solid boundary component |
| 21 | | Fifth Solid boundary component |
| 22 | | Sixth Solid boundary component |
| 23 | | Seventh Solid boundary component |
| 24 | | Eighth Solid boundary component |
| 25 | $5^2$ | 5-Space/Hyper cube-5 |
| 26 | | First Hyper Solid boundary component |
| 27 | | Second Hyper Solid boundary component |
| 28 | | Third Hyper Solid boundary component |
| 29 | | Fourth Hyper Solid boundary component |
| 30 | | Fifth Hyper Solid boundary component |
| 31 | | Sixth Hyper Solid boundary component |
| 32 | | Seventh Hyper Solid boundary component |
| 33 | | Eighth Hyper Solid boundary component |
| 34 | | Ninth Hyper Solid boundary component |
| 35 | | Tenth Hyper Solid boundary component |
| 36 | $6^2$ | 6-Space/Hyper cube-6 |

❑ ❑ ❑

# 43
# HYPER CUBES SEQUENCES-2 AND 3

Structural steps

Artifices formats-3

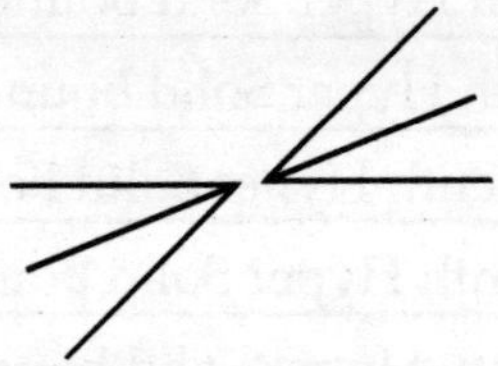

| | | | | | |
|---|---|---|---|---|---|
| 0 | 0 | $0^3$ | 0 | 0 | $0^3$ |
| –1 | –1 | $(-1)^3$ | 1 | 1 | $1^3$ |
| –2 | –2, –1<br>–3, –2 | $(-2)^3$ | 2 | 1, 2<br>2, 3 | $2^3$ |
| –3 | –3, –2, –1<br>–4, –3, –2<br>–5, –4, –3 | $(-3)^3$ | 3 | 1, 2, 3<br>2, 3, 4<br>3, 4, 5 | $3^3$ |
| 4 | –4, –3, –2, –1<br>–5, –4, –3, –2<br>–6, –5, –4, –3<br>–7, –6, –5, –4, | $(-4)^3$ | 4 | 1, 2, 3, 4<br>2, 3, 4, 5<br>3, 4, 5, 6<br>4, 5, 6, 7 | $4^3$ |

❑ ❑ ❑

# 44

# HYPER CUBES SEQUENCES-2

| | | | |
|---|---|---|---|
| 1 | $1^3$ | $6 \times 1 = 6 \times (0+1)$ | $1^3 + 6 \times (0+1)+1=2^3$ |
| 2 | | First half dimension | |
| 3 | | Second half dimension | |
| 4 | | Third half dimension | |
| 5 | | Fourth half dimension | |
| 6 | | Fifth half dimension | |
| 7 | | Sixth half dimension | |
| 8 | $2^3$ | | $2^3 + 6 \times (1+2)+1=3^3$ |
| gap | gap | $6 \times 3 = 6 \times (1+2)$ | |
| 27 | $3^3$ | | $3^3 + 6 \times (1+2+3)+1=4^3$ |
| gap | gap | $6 \times 6 = 6 \times (1+2+3)$ | |
| 64 | $4^3$ | | $4^3 + 6 \times (1+2+3+4)+1=5^3$ |
| gap | gap | $6 \times 10 = 6 \times (1+2+3+4)$ | |
| 125 | $5^3$ | | $5^3 + 6 \times (1+2+3+4+5)+1=6^3$ |
| gap | gap | $6 \times 15 = 6 \times (1+2+3+4+5)$ | |
| 216 | $6^3$ | | $6^3 + 6 \times (1+2+3+4+5+6)+1 =7^3$ |
| gap | gap | $6 \times 21 = 6 \times (1+2+3+4+5+6)$ | |
| 343 | $7^3$ | | $7^3 + 6 \times (1 + 2 + 3 + 4 + 5 + 6 + 7) + 1 = 8^3$ |
| gap | gap | $6 \times 28 = 6 \times (1+2+3+4+5 +6+7)$ | |

❑ ❑ ❑

# 45

# HYPER CUBES SEQUENCES-4

General Rule of Summation during synthesis of dimensions.

1. Let Dimension be N. And N – 2 be the dimension of dimension.
2. Synthesis of 2 dimensions.

   = Dimension + Dimension – (Dimension of Dimension)

   $= N + N - (N - 2) = N + 2$
3. Synthesis of 3 dimensions

   = Summation of (3–1) dimensions synthesis + dimension – (3-1) (Dimension of Dimension)

   $= (N + 2) + N - 2(N - 2) = 6$
4. Synthesis of 4 dimensions

   = Summation of (4 – 1) dimensions synthesis + dimension – (4 – 1) (Dimension of Dimension)

   $= (6) + N - 3(N - 2) = 12 - 2N = 2(6 - N)$
5. Synthesis of 5 dimensions

   = Summation of (5 – 1) dimensions synthesis + dimension – (5 – 1) (Dimension of Dimension)

   $= (12 - 2N) + N - 4(N - 2) = 20 - 5N = 5(4 - N)$

## 1. General Rule

6. Synthesis of R dimensions, $R > 1$

= Summation of (R – 1) dimensions synthesis + dimension– (R – 1) (Dimension of Dimension)

7. Illustratively for R = 2, synthesis rule would yield the summation value = Dimension + Dimension – Dimension of Dimension
8. For R = 2 and Dimension = N and Dimension of Dimension = N – 2, the above rule would yield summation value as N + N–(N – 2) = N + 2.
9. For R = 3 and Dimension = N and Dimension of Dimension = N – 2, the above rule would yield summation value as (N + 2) + N–(2N – 2)=6

| Ds | 1 Sp | 2 Sp | 3 Sp | 4 Sp | 5 Sp | 6 Sp |
|---|---|---|---|---|---|---|
| 1 | –1 | 0 | 1 | 2 | 3 | 4 |
| 2 | (–1)+( –1) – (–3) =1 | –2 | (1)+(1) – (–1) =3 | (2)+(2) – (0) =4 | (3)+(3) – (1) =5 | (4)+(4) – (2) =6 |
| 3 | 1+(–1) –2(–3) =6 | –4 | 3+(1) –2(–1) =6 | 4+(2) –2(0) =6 | 5+(3) –2(1) =6 | 6+(4) –2(2) =6 |
| 4 | 6+(–1) –3(–3) =–4 | –6 | 6+(1) –3(1) =4 | 6+(2) –3(0) =8 | 6+(3) –3(1) =6 | 6+(4) –3(2) =4 |
| 5 | (–4)+( –1) –4 (–3)= –15 $=1^4–2^4$=–15 | –8 | (4)+(1) –4 (1) =1 | (8)+(2) –4 (0) =10 | (6)+(3) –4(1) =5 | (4)+(4) –4 (2) =0 |
| 6 | (–15)+( –1) –5(–3)= –31 $=1^5–2^5$=–31 | –10 | 1+1 –5(1) =–3 | 10+2 –5(0) =12 | 5+3 –5(1) =3 | 0+4 –5(2) =–6 |
| 7 | (–31)+( –1) –6(–3)= –50 $=2^4–2^6$ | –12 | (–3)+(1) –6(1) =–8 | (12)+(2) –6(0) =14 | (3)+(3) –6(1) =0 | (–6)+4 –6(2) =–14 |
| 8 | (–50)+( –1) –7(–3)= –72 $=–2^3–2^6$ | –14 | (–8)+(1) –7(1) =–14 | (14)+(2) –7(0) =16 | (0)+(3) –7(1) =-4 | (–14)+4 –7(2) =–24 |
| 9 | (–72)+( –1) –8(–3)= –97 $=–1^5+2^5–2^7$ | –16 | (–14)+(1) –8(1) =–21 | (16)+(2) –8(0) =18 | (–4)+(3) –8(1) =-9 | (–24)+4 –8(2) =–36 |

General Rule of Summation during synthesis of dimensions :

1. Let Dimension be N. And N – 2 be the dimension of dimension.
2. Synthesis of 2 dimensions

   = Dimension + Dimension – (Dimension of Dimension)

   = N + N – (N – 2) = N + 2
3. Synthesis of 3 dimensions

   = Summation of (3 – 1) dimensions synthesis +dimension– (3 – 1) (Dimension of Dimension)

   = (N + 2) + N – 2 (N – 2) = 6
4. Synthesis of 4 dimensions

   = Summation of (4 – 1) dimensions synthesis +dimension – (4 – 1) (Dimension of Dimension)

   = (6) + N – 3(N – 2) = 12 – 2N = 2(6 – N)
5. Synthesis of 5 dimensions

   = Summation of (5–1) dimensions synthesis + dimension – (5–1) (Dimension of Dimension).

   = (12 – 2N) + N – 4(N – 2) = 20 – 5N = 5(4 – N)

## 2. General Rule

6. Synthesis of R dimensions, R > 1

   = Summation of (R–1) dimensions synthesis + dimension– (R–1) (Dimension of Dimension)
7. Illustratively for R = 2, synthesis rule would yield the summation value = Dimension + Dimension – Dimension of Dimension
8. For R = 2 and Dimension = N and Dimension of Dimension = N – 2, the above rule would yield summation value as N + N –(N – 2) = N + 2

❑ ❑ ❑

# 46

# DECIMAL CONVERSIONS OF FRACTIONS 1/9, 1/19, 1/29, 1/39, 1/49, ...

## 1. Double Digit Numbers

1. Double digit numbers 01 to 99 of ten place value system permit re-organization along 9 × 11 matrix format of 9 columns and 11 rows.
2. This availability of 9 × 11 matrix format for ten place value system, is a particular case of general rule for n place value system permitting re-organization for its double digits numbers in (n – 1) × (n = 1) matrix format of (n – 1) columns and (n = 1) row.
3. This as such would imply that n place value system shall be having (n – 1) numerals.
4. As in case of ten place value system, (10 – 1 = 9) numerals (1, 2, 3, 4, 5, 6,7, 8, 9) shall be of sequential values of rule (one more than before).
5. As such, say n = 19 shall be leading to 19 place value system with 18 numerals (1, 2, 3, 4, 5, 6,7, 8, 9, 10, 11, 12, 13, 14, 15, 16, 17, 18).
6. It is because of it that the fraction 1/19 shall be of the format of recurring decimal expression of 18 digits range. Likewise 1/29 shall be of the format of recurring decimal range of (29 – 1) = 28 digits range and ahead 1/39 to be of 38 digits decimal recurrence range, and so on.

## 2. Ten as Base and its Multiple Basis

7. With ten as base, $20 = 2 \times 10$ will be a multiple base with 2 as multiplier. Likewise $30 = 3 \times 10$ would be a multiple base 3 as multiplier, and so on.
8. Therefore, further as $10 - 1 = 9$, $20 - 1 = 19$, $30 - 1 = 29$ and so on, it shall be helping to reach at the decimal expressions values for fractions 1/19, 1/29, 1/39 and so on. With the help of their respective base multipliers namely 2, 3, 4 and so on.

## 3. Decimal Expression for 1/19

9. With 19 as place value, it shall be having a range of 18 numerals.
10. Further as that the multiple base $20 = 2 \times 10$ shall be providing 2 as a multiplier.
11. The fraction 1/9 with 9 as place value shall be having a range of 8 numerals. And with 10 as base, it shall be having $10 = 1 \times 10$, i.e. 1 as a multiplier.
12. With first numeral as '1', and multiplier being also '1', as such 8 numerals range, as such shall be of values '1, 1, 1, 1, 1, 1, 1, 1'.
13. With it the decimal expression value for 1/9 = 'decimal 1, 1, 1, 1, 1, 1, 1, 1', '1, 1, 1, 1, 1, 1, 1, 1' and so on.
14. It would lead to the repetition / reoccurrence for the range '1, 1, 1, 1, 1, 1, 1, 1'.
15. However as the range '1, 1, 1, 1, 1, 1, 1, 1' is also repetitive /recurrence value being '1', as such the decimal expression value for 1/9 is to be as $(.\overset{0}{1})$.
16. With further application of the rule of symmetry/ proportions of Ganita upsutra –1 and of 'one more than the previous one' of sutra –1, the decimal expression for 1/19 can be reached at from the decimal expression of 1/ 9 by applying 2 as multiplier for 18 steps range with sequential multiplication for the first numeral '1', which shall be leading us to:

(*i*) 1, (*ii*) $1 \times 2 = 2$,

(*iii*) $2 \times 2 = 4$, (*iv*) $4 \times 2 = 8$,

(*v*) $8 \times 2 = 16 = 10 = 6 = 6$ with 1 as carry forward value,

(*vi*) $6 \times 2 = 1 = 13$, 3 with 1 as carry forward,

(*vii*) $3 \times 2 = 1 = 7$,

(*viii*) $7 \times 2 = 14 = 4$ with 1 as carry forward,

(*ix*) $4 \times 2 = 1 = 9$,

(*x*) $9 \times 2 = 18 = 8$ with 1 as carry forward,

(*xi*) $8 \times 2 = 1 = 17$ 7 with 1 as carry forward,

(*xii*) $7 \times 2 = 1 = 15 = 5$ with 1 as carry forward,

(*xiii*) $5 \times 2 = 1 = 11 = 1$ with 1 as carry forward,

(*xiv*) $1 \times 2 = 1 = 3$,

(*xv*) $3 \times 2 = 6$,

(*xvi*) $6 \times 2 = 12 = 2$ with 1 as carry forward,

(*xvii*) $2 \times 2 = 1 = 5$,

(*xviii*) $5 \times 2 = 10 = 0$ with 1 as carry forward.

**Note:** – Reaching at '0' as digit value and '1' as carry forward, as such shall be leading to $0 \times 2 = 1 = 1$, i.e. the same as the value as it was at the initial stage of 'i' step above, and as such the process would repeat and would continue repeating and hence the decimal recurrence value expression for 1/19 being as follows:

$1/19 = .\overset{8}{0}526315789473684 2.$

17. As in reference to 10, the value 20 is reached at with the help of 2 as a multiplier of 10, so in the reverse order value 10 can be attained with the help of 2 as a divisor of 20.
18. Therefore the decimal expression for 1/19 can be attained by sequential division of 1 by 2 for 18 times.
19. Further, It would be relevant to note that as 18 numerals range makes out 9 reflection pairs of complementary ranges for the biggest numeral of ten place value system,

as such the working uptill nine steps, i.e. up till half of the range of 18 numerals, and thereafter the second half range of steps 10 to 18 to be computed respectively as of complementary values for the first nine steps digits.

20. For facility of convenience these features may be tabulated by arranging the values of 18 digits expression range as follows.

| ix | viii | vii | Vi | v | iv | iii | ii | i |
|---|---|---|---|---|---|---|---|---|
| 9 | 4 | 7 | 3 | 6 | 8 | 4 | 2 | 1 |
| 0 | 5 | 2 | 6 | 3 | 1 | 5 | 7 | 8 |
| xviii | xvii | xvi | xv | xiv | xiii | xii | xi | X |
| 9 | 9 | 9 | 9 | 9 | 9 | 9 | 9 | 9 |

## 4. Decimal Expression Format for 1/ 39

21. For 1/39, as 39 – 1 = 38, as such numerals range for 39 place value shall be of the 38 digits range. Further as 40 = 4 × 10, as such the base multiplier in reference to base 10 shall be '4'. Therefore the first numeral value '1' is to be sequentially multiplied by 4 to

    The recurring decimal expression for 1/39

**Note:** As 39 = 3 × 13, as such the above recurrence range format of 38 digits shall be reduced because of its internal osculation because of '13' being the factor.

## 5. Decimal Expression for 1/49, 1/59,—

22. It would be blissful exercise to reach at recurring decimal expressions values for 1/ 49, 1/ 59 and so on , with the help of 5, 6 and so on multipliers for 48, 58 and so on numerals digits ranges.

❑ ❑ ❑

# 47

# AUXILIARY FRACTIONS

1. Vedic systems have the tool of auxiliary fractions. This tool makes it very convenient to reach at decimal expressions for the fractions.
2. Of these, the most glaring illustration is of the fractions of the type 1/19, 1/29, 1/39 and so on where the last digit of the denominator is '9'. From the following table of auxiliary fraction for these fractions, it may the would be glaringly visible the practical use of the auxiliary fractions to reach at the decimal fractions for the concerned expression:

| Fraction | Auxiliary fractions |
|---|---|
| 1/19 | 1/2 |
| 1/29 | 1/3 |
| 1/39 | 1/4 |
| 1/49 | 1/5 |
| 1/59 | 1/6 |
| 1/69 | 1/7 |
| 1/79 | 1/8 |
| 1/89 | 1/9 |

1. **Further as 1 × 9 = 9, 3 × 3 = 9 and 7 × 7 = 9 as such the denominators with last digit as '1', the multiplication of**

the numerator and denominator by '9' will reduce the denominator with last digit as '1 × 9' = 9. Likewise the denominator to the last digit as '3' by multiplication of the numerator and denominator by '3' shall be reducing it as to be of last digit as '3 × 3' = 9. Likewise the denominator with last digit as 7 can be reduced with last digit as '9' by multipliying the numerator, as well as the denominator with 7.

2. Accordingly all the denominators with last digit as '1' or '3', or 7 or 9 would get reduced into the denominator with last digit as '9' by suitable multiplications of the numerators and denominators availing the features 1 × 9 = 9, 3 × 3 = 9, 7 × 7 = 9 and 9 × 1 = 9.
3. The denominators with last digits as '0, 2, 4, 5, 6 and 8' can be first reduced by taking out factors as '2' or '5'. This process with repeated steps shall be ultimately leading to one of the factors of the denominators being of last digit as '1 or 3, 7 or 9'. Therefore every denominator, ultimately would get reduced to factors with last digits being '2, 5 or 9'.
4. As the division by 2 or 5 or 9 is simple by vedic systems, as such, the expression for fractions in decimal form would be a pleasant exercise.
5. Here is being tabulated the decimal expression for the fractions 1/ n, n = 2, 3, 4 , — 20.

| 1/1 | 1 |
|---|---|
| ½ | .5 |
| 1/3 | $.\dot{3}$ |
| ¼ | 1/2 × 1/2 |
| 1/5 | .2 |
| 1/6 | 1/2 x 1/3 |
| 1/7 | $.\dot{1}4285\dot{7}$ |
| 1/8 | 1/2 × 1/4 |

| | |
|---|---|
| 1/9 | 1/3 × 1/3 |
| 1/10 | 1/2 × 1/5 |
| 1/11 | $0.\dot{0}\dot{9}$ |
| 1/12 | 0.08 |
| 1/13 | $0.\dot{0}7692\dot{3}$ |
| 1/14 | $0.0\dot{7}1428\dot{5}$ |
| 1/15 | $0.0\dot{6}$ |
| 1/16 | 0.0625 |
| 1/17 | $0.\dot{0}58823529411764\dot{7}$ |
| 1/18 | $0.0\dot{5}$ |
| 1/19 | $0.\dot{0}5263157894736842\dot{1}$ |
| 1/20 | 0.05 |
| 1/21 | $0.\dot{0}4761\dot{9}$ |
| 1/22 | $0.0\dot{4}\dot{5}$ |
| 1/23 | $0.\dot{0}43478260869565217391\dot{3}$ |
| 1/24 | $0.041\dot{6}$ |
| 1/25 | 0.04 |
| 1/26 | $0.0\dot{3}8461\dot{5}$ |
| 1/27 | $0.\dot{0}3\dot{7}$ |
| 1/28 | $0.0357\dot{1}4285\dot{7}$ |
| 1/29 | $0.\dot{0}344827586206896551724137593\dot{1}$ |
| 1/30 | $0.0\dot{3}$ |
| 1/31 | $0.\dot{0}3225806451612\dot{9}$ |

| | |
|---|---|
| 1/32 | 0.03125 |
| 1/33 | $0.0\dot{3}\dot{0}$ |
| 1/34 | $0.0\dot{2}94117647058823\dot{5}$ |
| 1/35 | $0.0\dot{2}8571\dot{4}$ |
| 1/36 | $0.02\dot{7}$ |
| 1/37 | $0.\dot{0}2\dot{7}$ |
| 1/38 | $0.0\dot{2}6315789473684210\dot{5}$ |
| 1/39 | $0.\dot{0}2564102564\dot{1}$ |
| 1/40 | 0.025 |
| 1/41 | $0.\dot{0}243\dot{9}$ |
| 1/42 | $0.0\dot{2}3809\dot{5}$ |
| 1/43 | $0.\dot{0}232558139534883720 9\dot{3}$ |
| 1/44 | $0.02\dot{2}\dot{7}$ |
| 1/45 | $0.0\dot{2}$ |
| 1/46 | $0.0\dot{2}17391304347826086956\dot{7}$ |
| 1/47 | $0.\dot{0}2127659574468085106382$<br>$9787234043553191489361\dot{7}$ |
| 1/48 | $0.0208\dot{3}$ |
| 1/49 | $0.\dot{0}20\ 408\ 163\ 265\ 306\ 122\ 448\ 979\ 591\ 836\ 734$<br>$693\ 877\ 55\dot{1}$ |
| 1/50 | 0.02 |
| 1/51 | $0.\dot{0}19607843137254\dot{9}$ |
| 1/52 | $0.0\dot{9}12307\dot{6}$ |

| | |
|---|---|
| 1/53 | 0.$\dot{0}$18867924528$\dot{3}$ |
| 1/54 | 0.0$\dot{1}$8$\dot{5}$ |
| 1/55 | 0.0$\dot{1}$$\dot{8}$ |
| 1/56 | 0.017857$\dot{1}$4285$\dot{7}$ |
| 1/57 | 0.$\dot{0}$17543859649122807$\dot{7}$ |
| 1/58 | 0.0$\dot{1}$72413793103448275862068965$\dot{7}$ |
| 1/59 | 0.$\dot{0}$169 4915 2542 3728 8135 5932 2033 8<br>9830 5084 7457 6271 1864 4067 7966$\dot{1}$ |
| 1/60 | 0.01$\dot{6}$ |
| 1/61 | 0.016393442622950819672131147540984 |
| 1/62 | 0.0$\dot{1}$6129032258064$\dot{5}$ |
| 1/63 | 0.$\dot{0}$1587$\dot{3}$ |
| 1/64 | 0.015625 |
| 1/65 | 0.0$\dot{1}$5384$\dot{6}$ |
| 1/66 | 0.0$\dot{1}$$\dot{5}$ |
| 1/67 | 0.014925373134328358208955223880596 |
| 1/68 | 0.01$\dot{4}$70588235294117$\dot{6}$ |
| 1/69 | 0.$\dot{0}$144927536231884057971$\dot{1}$ |
| 1/70 | 0.0$\dot{1}$4285$\dot{7}$ |
| 1/71 | 0.0140845070422535211267605633380282 |
| 1/72 | 0.013$\dot{8}$ |
| 1/73 | 0.$\dot{0}$136986$\dot{3}$ |
| 1/74 | 0.0$\dot{1}$3$\dot{5}$ |
| 1/75 | 0.01$\dot{3}$ |
| 1/76 | 0.01$\dot{3}$157894736842105$\dot{2}$6 |

| | |
|---|---|
| 1/77 | $0.\dot{0}1298\dot{7}$ |
| 1/78 | $0.0\dot{1}2820\dot{5}$ |
| 1/79 | $0.\dot{0}12658227848\dot{1}$ |
| 1/80 | 0.0125 |
| 1/81 | $0.\dot{0}1234567\dot{9}$ |
| 1/82 | $0.0\dot{1}219\dot{5}$ |
| 1/83 | 0.012048192771084337349397590361446 |
| 1/84 | $0.01\dot{1}9047\dot{6}$ |
| 1/85 | $0.0\dot{1}17647058823529\dot{4}$ |
| 1/86 | $0.0\dot{1}16279069767441860046\dot{5}$ |
| 1/87 | $0.\dot{0}11494252873563218390804597\dot{9}$ |
| 1/88 | $0.011\dot{3}\dot{6}$ |
| 1/89 | 0.011235955056179775280898876404494 |
| 1/90 | $0.0\dot{1}$ |
| 1/91 | $0.\dot{0}1098\dot{9}$ |
| 1/92 | $0.01\dot{0}86956521739130434782\dot{6}$ |
| 1/93 | $0.\dot{0}1075268817204\dot{3}$ |
| 1/94 | 0.010638297872340425531914893617021 |
| 1/95 | $0.\dot{0}0052631578947368 4\dot{9}$ |
| 1/96 | $0.01041\dot{6}$ |
| 1/97 | 0.010309278350515463917525773195876 |
| 1/98 | $0.\dot{0}1020408163265306122448979591836734693$<br>$877\dot{5}$ |
| 1/99 | $0.\dot{0}\dot{1}$ |
| 1/100 | 0.01 |
| 1/101 | $0.\dot{0}09\dot{9}$ |
| 1/102 | $0.0\dot{0}98039215686274\dot{5}$ |

| | |
|---|---|
| 1/103 | 0.009708737864077<br>6699029126213592233 |
| 1/104 | $0.009\dot{6}1538\dot{4}$ |
| 1/105 | $0.0\dot{0}9523\dot{8}$ |
| 1/106 | $0.0\dot{0}94339622641\dot{5}$ |
| 1/107 | 0.0093457943925233<br>6448598130841121 5 |
| 1/108 | $0.0\dot{0}02\dot{5}$ |
| 1/109 | 0.0091743119266055<br>045871559633027523 |
| 1/110 | $0.0\dot{0}\dot{9}$ |
| 1/111 | $0.\dot{0}2\dot{7}$ |
| 1/112 | $0.0089\dot{2}8571\dot{4}$ |
| 1/113 | 0.008849557522123<br>893805309734513274 3 |
| 1/114 | $0.0\dot{0}8771929824561403\dot{5}$ |
| 1/115 | $0.0\dot{0}869565217391304347826\dot{6}$ |
| 1/116 | $0.00\dot{8}62068965517241379310344827\dot{5}$ |
| 1/117 | $0.\dot{0}0854\dot{7}$ |
| 1/118 | 0.0084745762711864<br>406779661016949153 |
| 1/119 | 0.0084033613445378<br>151260504201680672 |
| 1/120 | $0.008\dot{3}$ |

# 48

# FURTHER REVISING GANITA SUTRA-1

## 1. Ganita Sutras

1. Ganita Sutras together with Ganita upsutras is a complete scripture.
2. It being a vedic scriptures, its start with source reservoir is the (sole syllable Braham) / Om (ॐ) and its end fruit is Its synonym /Parnav प्रणवः.
3. Being a complete Vedic scripture, it is to approached in terms of established two fold processing process, namely 'Sankhiya Nishtha' and 'Yoga Nishtha'.

## 2. Established Processing Process Approach

1. Two fold established processing process approach is of 'Sankhiya Nishtha' and 'Yoga Nishtha' complementing and supplementing each other.
2. The Sankhiya Nishtha on its ultimate analysis presumes existence of geometric formats for working along artifices of numbers.
3. Yoga Nishtha, on the other hand presumes existence of artifices of numbers and works out the geometric formats as dimensional spaces.
4. As such, it is the interrelationship of artifices of numbers and of dimensional frames which are availed for their

applied values and these interrelationships become the subject matter of chase of mathematics as the discipline of the pure values and virtues.

## 3. Chase of Pure Values and Virutes

5. Vedic mathematics, as such, may be accepted by way of definition being the discipline of chase of pure values and virtues of the existence phenomenon provided by interrelationship of artifices of numbers and of dimensional frames of spaces.

## 4. Start with Source Reservoir

6. The startwith source reservoir of values and virtues, as per the enlightenment of ancient wisdom with Vedas as core, is the 'sole syllable Braham'/'Om (ॐ)'.
7. The values and virtues of the existence phenomenon are accepted as flowing through the four components formulation of Om (ॐ) starting with Bindu Sarovar and reaching up till Swastik pada through Ardh matra and Tripundum, as a Divya Ganga Flow along artifices of numbers this flow is of nine streams seven streams, from the above and is of single stream and three stream from below as a two fold flow within transcendental domain.

## 5. Ganita Sutra-1

8. Ganita Sutra-1 'ekadhiken purvena' (एकाधिकेन पूर्वेण) is a composition of two formulations namely (1) एकाधिकेन and (2) पूर्वेण.
9. This pair of formulation namely (1) एकाधिकेन and (2) पूर्वेण are composition of nine letters and seven letters respectively.
10. This structural set-up of the text of Ganita Sutra-1 as of a pair of formulation of nine letters and seven letters, as such avails the pair of artifices (9, 7) and thereby there is a jump over the in between artifice value (8).

11. It is this jump/gap, which as such becomes the focus of the structural message.
12. This as such deserves to be comprehended and imbibed as a 'feature' / structural feature.
13. The pure value and virtue of this feature may be viewed as transcendence through this gap/artifice 8/Asht Prakrati.
14. It is the transcendence of the Brahman values and virtues/artifice 9/9-space/Nav Braham through Asht Prakrati/eight fold nature and manifestation ahead as unity state of consciousness/Sapt Rishi lok/7-space/ artifice 7.
15. It also may be viewed as that Asht Prakrati/eight fold nature causes a gap which is needed to be bridged for the unity state of consciousness to be in unison with the values and virtues of Nav Braham.
16. One may have a pause here and have a fresh look at artifice 8 permitting re-organization as $2 \times 2 \times 2 = 2^3$. It is this re-organization which shall be sequentially taking from half dimension to quarter of pair of dimensions to octave cut of all the three dimensions of linear order 3-space set up and hence getting blocked at the upper limit for the linear order set up.
17. Here it would be relevant to note that from 9 points fixation of a cube to its 7 versions range, the in between phase and stage is going to be of the features of split up of the cube into eight sub cubes.
18. This as such shall be bringing to focus as that the pure and applied values of Ganita Sutra-1 are going to be multi fold.
19. One such feature is going to be as that ahead of $2^3$ would be $2^4 = 16$ which is the combined range of pair of formulations of the text of Ganita Sutra-1 $(9 + 7 = 16)$.
20. One another feature of this split of the text into a pair of formulations with one gap shall be helping comprehend

and imbibe the values of a working rule of Ganita Sutra-1 being by 'one more than the previous one'.

21. Still further, the pair (9, 7) shall be permitting re-organization as (8 + 1), (8 – 1).
22. The organization (8 + 1), (8 – 1) shall be leading to the organization format 7 × 9 matrix of 7 columns and 9 rows.
23. This as such, infact is going to be the organization format for double digit numbers of eight place value system as follows:

| | | | | | | | |
|---|---|---|---|---|---|---|---|
| 01 | 02 | 03 | 04 | 05 | 06 | 07 | |
| 10 | 11 | 12 | 13 | 14 | 15 | 16 | |
| 17 | 20 | 21 | 22 | 23 | 24 | 25 | |
| 26 | 27 | 30 | 31 | 32 | 33 | 34 | |
| 35 | 36 | 37 | 40 | 41 | 42 | 43 | |
| 44 | 45 | 46 | 47 | 50 | 51 | 52 | |
| 53 | 54 | 55 | 56 | 57 | 60 | 61 | |
| 62 | 63 | 64 | 65 | 66 | 67 | 70 | |
| 71 | 72 | 73 | 74 | 75 | 76 | 77 | 80 |

24. One may have a pause here and have a fresh look at the text of Ganita Sutra-1 and Ganita Sutra-2.
25. Ganita Sutra-1 text is of the pair of formulations (1) एकाधिकेन and (2) पूर्वेण with one gap and Ganita Sutra-2 text is of three formulations (1) निखिलं (2) नवतष्चरमं (3) दशतः with a pair of gaps.
26. The working rule of Ganita Sutra-2 'all from 9 and last from 10', leads to 'ten place value system'.
27. The transition from Ganita Sutra-1 leading to 'eight place value system' to 'ten place value system', with a jump over 'nine place value system', in a way would help have an insight about the working rules of Ganita Sutra-1 as well as of Ganita Sutra-2.

❑ ❑ ❑

# 49

# A WORD WITH THE TEACHERS

Young minds at impressionable stages, naturally are to be instructed very gently particularly so when the conceptual base of basics along first principles is being aimed.

Presuming informal acquaintance with the counting numbers, the formal mathematical base is to be formatted step by step. Pebbles and objects based counting, as a starting point, a smooth transition deserves to be attained by associating points along the line as the seats of the pebbles.

Till the comprehension of complete association of counting pebbles with their placement points along the line is fully stands grasped and imbibed by the child, the experiment is continued to be repeated.

The movement this association of counting pebbles with the corresponding point along the line as seats of the respective pebbles would be part of the cherishable memory of the child, it be taken as the blissful step of arousing a promising mathematical line of thinking for the young mind.

It may be taken as the face and stage of introducing mathematical tools. The introduction of scale and exposure to its making as having association of counting numbers with the points along a line will go a long way to add to the mathematical line of thinking as to be to the values like that of units for distance parallel to the values of the counting numbers.

The values as of counting numbers and of units of scale along a line from a given reference point along the line would help introduced. Reverse counting availing the orientations of the line. It would be like forward moving and backward moving exercises. In the background of forward moving steps the increasing values feature of the counting numbers may be focused. The arithmetic operation of addition as well can be formally introduced and demonstrated.

Parallel to it, with the help of backward moving steps. The decreasing values feature of the reverse counting may be focused. Here the arithmetic operation of subtraction can be very well introduced and demonstrated.

With conceptual comprehension of addition and subtraction operations as independent operations a phase and stage would emerge for smoothly introducing the interconnected features of addition and subtraction operations.

Here opportunity can be availed to very gently introduced the absolute value of counting numbers as distinguishable from positive and negative values of the counting numbers.

Once the young mind shall be feeling confident in handling direct counting with the help of the working rule (one more than the previous one, and reverse counting in terms of the working rule 'one less than the previous one', it may be taken that the mathematical foundation stands laid down for the thinking child.

It would be a phase and stage of formal teaching of mathematics for young minds with the help of Ganita Sutras.

❑ ❑ ❑

# 50

# MENTAL MATHEMATICS

The basic features of Ganita Sutras Vedic mathematics is being the mental mathematics. The basics on first principles of whole range of mathematics have been crystallized as sutras (hymns) which are short and simple.

There are just sixteen Sutras and thirteen upsutras in all. The entire sutras text (including that of Upsutras) is availing only 520 letters, and as such it is easy to memorize and to dispense with the written text all together.

It is this feature of the Ganita Sutras text, which easily makes it possible to memorize and to dispense with it all together, which makes it a mental mathematics as that everything gets printed in mind itself.

It is not only that the text is short and can be memorized, but also that the basics as well are ultimately reduced to the need of memorizing the arithmetic operations only up till '5 × 5, 5 multiplied by 5 results' to handle everything else simply as mental operations, which makes this approach of Vedic mathematics, the mental mathematics.

Further, though, there are sixteen Sutras and thirteen upsutras but they as well are nothing but the expended elaborations of the text and working rule of Ganita Sutra-1 and that being so, ultimately the whole thing would get reduce to the understanding and imbibing of the values of the sixteen letters text and working rule of Ganita Sutra-1.

It would be a blissful exercise to approach the text of Ganita Sutras, in the sequence and order of the Sutras and Upsutras. The text of each Sutra (and of each upsutra) deserves to be approached in the sequence and order of the letters availed by the text of the Sutra (Upsutra). It would bring one face to face with the internal structure of the texts of the Sutras. This friendship with the internal structural compositions of the texts of the Sutras and Upsutras shall be going a long way, not only to help memorize the text, but also to have insight about their working rules and as an end result to help mathematize the thinking process itself as to be of the format of the basics of mathematics on first principles.

Though the initial expectations for exposing the young minds to the mathematical path of Ganita Sutra is of acquaintance with the arithmetic operations up till '5 × 5', but after being through the process for some steps, it would come to focus as that the real need for this mathematical path is proper comprehension and imbibing of the concept and values of 'one'. Therefore, to start with, it would be appropriate to mature the young minds about this concept and value of '1', to ensure that, two different objects, would permit each being associated with the value '1'.

❑ ❑ ❑

# 51

# NUMBER LINE

## 1. Introductory

There is a common pool of Ancient Wisdom with Ved as the core source thereof. Vedic knowledge traditions are to urge to know Ved in terms of Ved itself. The organized knowledge of Ved as four Vedas, Rig, Yaju, Sam and Atharav accepts knowing first three Vedas in terms of fourth Ved, Atharv Ved. The first quarter of first Mantra of Atharav Ved enlightens about the first principle of knowing Vedic knowledge. The Mantra, as its first quarter enlightens:

Yeh Trishapta Paryanti Vishwa

ये त्रिशप्ता पर्यन्ति विश्वः

This Trishapta envelops Vishwa.

"Trishapta/त्रिशप्ता" is a technical term, whose one prominent expression in terms of artifices of numbers is "3 and 7"; of which, the artifices coordinations organizations, come to be 37 and 73. There are other coordinations organizations of artifices of 3 and 7 as well like 3+7, 3/7, 7/3, 3×7, $3^7$, $7^3$, 3.7, 7.3 and so on. However, here at this initial stage of very first lesson of this introductory course of learning and teaching simultaneously for students and teachers together, the initial focus at initiation stage of Discipline of Mathematics, in terms of Numbers as tools as counts and formats for coordination and organization of artifices of numbers, the initial coordinations organization of artifices of 3 and 7 are being focused as 37 and 73.

This coordination organization focus of artifices of 3 and 7 as 37 and 73 is availed for organization of knowledge of "Sidha Sidhanta Padati" also known as Gorakshko Upanishad, the source scripture of Shiva cult sect, Gorakh Panthi. The scripture is of six Updesha (enlightenment sermons). The first Updesha is titled "Pinda Utpati/ emergence of body within human frame from within another body within human frame". The range of this Updesha is of 73 shalokas length. The middle shaloka, 37 Shaloka, enlightens about Asth Sakar Murti/hyper cube-4, representative regular body of creator's Space/4-Space enveloped within eight solid boundary components.

The enlightenment up till sixth Updesha is of the transcendental order of transcendence into Pursha domain/orb of Sun/Vishnu Lok/6-Space. Further Upanishadic enlightenment is that the Vishnu Lok/ 6-Space/Pursha format accepts 26 tatav/ basic elements. Sankhya approach to these basic elements is along the formats of the artifices of numbers 1 to 26. Initially it may sound all strange if stated as that English Alphabet of 26 letters (A to Z) talks mathematics of the coordinations organizations of artifices of numbers 1 to 26 parallel to letters A to Z but the studies of the author convince as that this is there because of fundamental unity of Ancient wisdom identically feeding from same course source all the frequencies which are accepted by human mind from generation to generation.

Pinda Utpati/emergence of body, is conceptually, about the phenomena of birth of a child. The number value format (NVF) for CHILD for the letters availed (C = 3, H = 8, I = 9, L = 12 and D = 4 total 36) and number value format (NVF) for CHILDREN for the letters availed (CHILD= 36, R=18, E=5, and N=14 total 73), as such takes us to the identical counts formats for the organizations of knowledge.

NVF (NUMBER) = NVF(COUNT) = NVF(FORMAT) = 73 and reflection pair of 73 is 37 and same is NVF(CONE). As such the title of the first lesson is settled as NUMBER CONE.

## Number Cone

Number as count, and as format, as artifice of 73 accepting organization as 36 +1+ 36, and further as $6 \times 6 + 1 \times 1 + 6 \times 6$

takes us to the spatial/plane/square/ 2-Space order. The only aspect being availed here for the initial stage of approaching numbers as counts is that the pairing with NVF (PAIRING) = 74 = 37 + 37, is being had such that we to have a pair of cones and then to proceed for coordination organization for the counts in terms of pair of boundary lines of spatial cut of a cone as under:

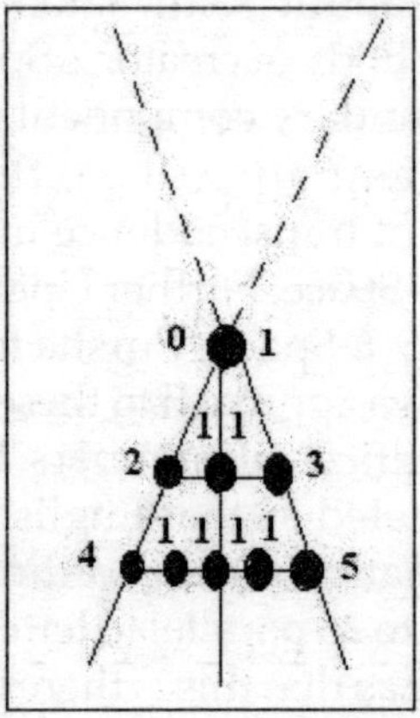


## 2. Lesson Steps

Stage-1: To help the students to draw above number-cone.

**Stage-2:** To help the students to familiarize with the depiction and placements of points as well as fixations of linear units along different steps of cone within its pair of boundary lines as spatial/plane/2-Space print out of the cone as above.

**Stage-3:** To specifically chase points counts and linear units count as

1. One point is one count.
2. Three points are three counts. Three points also fix a pair of units (linear units as counts).
3. Five points are five counts. Five points also fix quadruple units (linear four counts fixed by five points).

4. In general N points help fix (n–1) linear units

**Stage-4:** To teach counting chase along above number-cone format

**Stage-5:** Point out as that in this exposure of this approach to whole numbers set as counting numbers set special features of organization of counting numbers as whole numbers being focused are:

1. As that counting and table of 2 are being simultaneously formatted.
2. Counts can begin with zero as well as with one
3. Counts as counting set as whole numbers accepts partition as evens and odds.
4. Evens array has additional message as that it is also the table of 2.
5. There is a further interactive message of the odds array and even arrays that they can interact with each other by both rules of "one more" and "one less".
6. The odds array as well as evens array further convey that these are numbers lines with one jump in between its consecutive members placements.
7. The odds array can also give us 1, 1 × 3, 1 × 5.... While evens array can give us 2, 2 × 4, 2 × 4 × 6....

**Note:** The following stage-6 is being added just to feed the curiosity as to why the above seventh feature is being mentioned; the focus may be that like partition of whole numbers as odds and evens, the dimensional spaces as well accept parallel partition as odd dimensional spaces and even dimensional spaces and that ultimately the artifices of numbers are to run parallel to the dimensional frames of spaces.

**Stage-6:** It may be introduced as that odds array supply 1, 1 × 3, 1 × 5.... Shall be taking us to dimensional order ranges of odd dimensional spaces while the evens array supply 2, 2 × 4, 2 × 4 × 6....shall be taking us to two dimensional order ranges of even dimensional spaces

# 52

# APPLICATION OF WORKING RULE OF GANITA SUTRA-1 (one more than the previous one)

1. Construction of Whole Numbers

1. Let us presume, we know counting number '1'.

   Let us take it to be as our previously known number.

   Now let us apply the working rule of the Sutra,

   It shall be taking us from '1' to '1+1'=2.

2. It will make '2' as our previously known number.

   With application of the working rule of the Sutra,

   It shall be taking us from '2' to '2+1'= 3.

3. With it, '3' would be our previously known number,

   And application of the working rule shall be taking us from '3' to '3 + 1' = 4.

4. And like that the whole range of counting numbers can be reached at with the help of sequential application of the working rule of Ganita Sutra-1.

5. **Exercise:** To sequentially reach at counting numbers 5 to 10 with the help of the working rule of Ganita Sutra-1.

**Hint. First step:** Note down the previously known number.

**Step two:** Add previously known number and one.

**Step three:** Reach at the result.

❑ ❑ ❑

# 53

# APPLICATION OF GANITA SUTRA-1

## Construction of mathematical tools

### 1. To Construct A Scale

1. Point has no length, as such it can be represented by '0'
2. Let point (0) be the previous object (number)
3. With application of the Ganita Sutra-1 working rule, from point (0) reach would be to line (length) as 0 + 1 = 1
4. Draw a line
5. Fix point on the line as '0'
6. Fix another point on the line
7. Take the distance between these two points as a unit (1)
8. Add value to the starting point as '0' and to the next point as 0+1=1

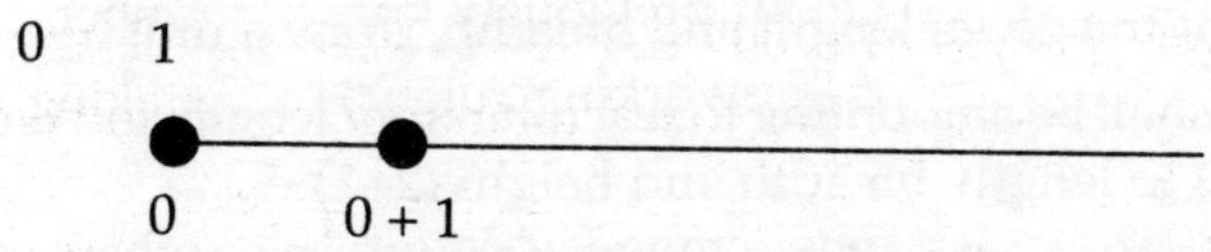


9. Fix another point on the line at a unit distance and associate it with the value '1+1=2'

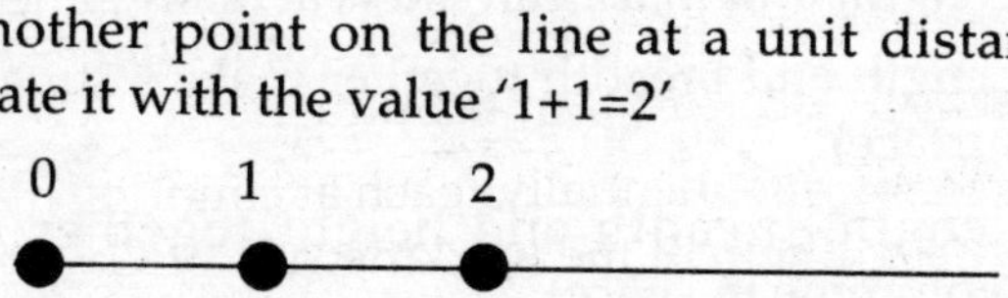


10. It would be a blissful exercise to fix further points on the line and to make it a scale of as many units as one likes.

❑ ❑ ❑

# 54

# LENGTH, BREADTH AND HEIGHT

1. To Construct an Interval (), Square and Cube

1. With the help of the scale, draw a line of unit length.
2. Take it as the previous number value (1).
3. Draw unit breadth for the given unit length.
4. It shall be amount to reaching from length (1) to Length and breadth (1+1)=2.
5. For the given length and breadth, draw a unit height.
6. It shall be amounting to reaching from length and breadth (2) to length, breadth and height (2+1)=3.
7. Length, unit makes an interval (along line).
8. Length and breadth together make a square (area along surface).
9. Length, breadth and height together make a cube (volumme in space).
10. It would be a blissful exercise to draw intervals of different units lengths.
11. It also would be a blissful exercise to draw squares of different units areas.

12. Further it would be a very blissful exercise to draw cubes of different volummes.

## 2. Construction of whole numbers

'1' is 'one more than the previous (0)'.

'2' is 'one more than the previous (1)'.

'3' is 'one more than the previous (2)'.

And so on.

## 3. Construction of Dimensional Frame

'Point' has no length and as such is of 'zero' dimension/axis.

'Line' has length and as such it has one dimension than before that is of (of point).

'Surface' has area and as such it has two dimensions/axes that is, one more than the previous (one dimension of length).

'Solid' has volumme and as such it has three dimensions/axes, that is one more than the previous (two dimension of surface/area'

## 4. Construction of Axes/dimensional Frames.

One axis It can be constructed with the help of ruler/scale.

It shall be of the features of the straight line.

It shall be representing the length of a line.

Two axes It can be constructed with the help of ruler and a set square.

It shall be of the features of the pair of crossing lines.

It shall be representing as length and breadth of surface area.

Three These three axes can be constructed as three axes emanating axes from the corner point of a cubeboid

Three axes shall be representing as length, breadth and height of solids.

## 5. Interval, Square and Cube

**Interval** Interval is a geometric body accepting one axis/one axis frame/one dimensional frame.

As such 'interval' would be a 1-space body.

**Square** Square is a geometric body accepting a pair of axes/ two axes frame/two dimensional frame.

As such square would be a 2-space body.

**Cube** Cube is a geometric body accepting three axes/three axes frame/three dimensional frame.

As such cube would be 3-space body.

## 6. Origin Point

**Origin point** Origin point formally may be defined and accepted as a point where axes cross.

The corner points of the cube would be the origins of three axes frames.

The corner points of the square would be the origin points of two axes frames.

The end points of the interval would be the origin points of the one axis frame.

❑ ❑ ❑

# 55
# DOMAIN BOUNDARY RATIO

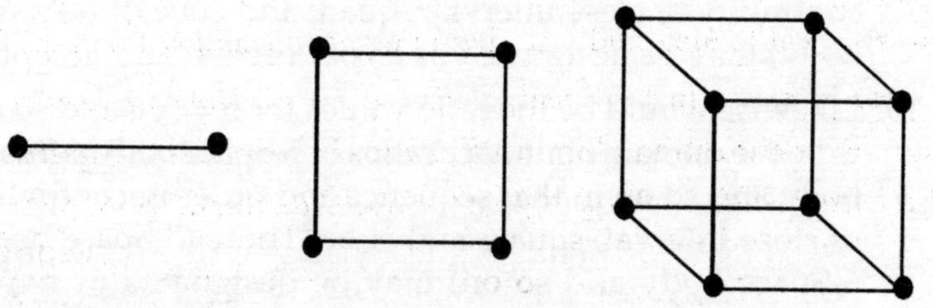

| Close Interval | Square | Cube |
|---|---|---|

1. Close interval gives us domain (length) as $A^1$ and the boundary (pair of end points) as $2A^0$.
2. This becomes the ratio $A^1 : 2A^0$.
3. Square gives us domain (area) as $A^2$ and the boundary as $4A^1$.
4. This becomes the ratio $A^2 : 4A^1$.
5. If we have a close look at the ratios

   (*i*) $A^1 : 2A^0$ and

   (*ii*) $A^2 : 4A^1$,

   we can decipher the pattern for these ratios as

   $A^n : 2nA^{n-1}$ for n = 1 and 2

6. Now if we put the value n = 3, we get the ratio

   (*iii*) $A^3 : 6A^2$.

7. This is precisely domain (volume) of cube as $A^3$ and boundary (six surfaces) of cube as $6A^2$.

8. Once the ratio $A^n: 2nA^{n-1}$ holds for n = 1, 2 and 3 for our known geometric bodies, close interval, square and cube as representative regular bodies of 1-space, 2-space and 3-space respectively, the query would be as to how this sequence for values n = 4, 5, 6 and for onward values would behave?

9. Natural answer would be that the ratio values for n = 4 is to be the domain boundary ratio of 4-Space body in continuity of close interval, square and cube. It (4-Space body) may be designated as hyper cube-4.

10. Likewise would be the ratio values for n = 5, 6 and so on to be the domain boundary ratios of 5-Space body, 6-Space body and so on in that sequence and order in continuity of close interval, square and cube. These (5-Space body, 6-Space body and so on) may be designated as hyper cubes 5, 6 and so on.

❑ ❑ ❑

# 56

# GEOMETRIC COMPONENT FORMULATION $(A + 2)^n$, n = 1,2,3

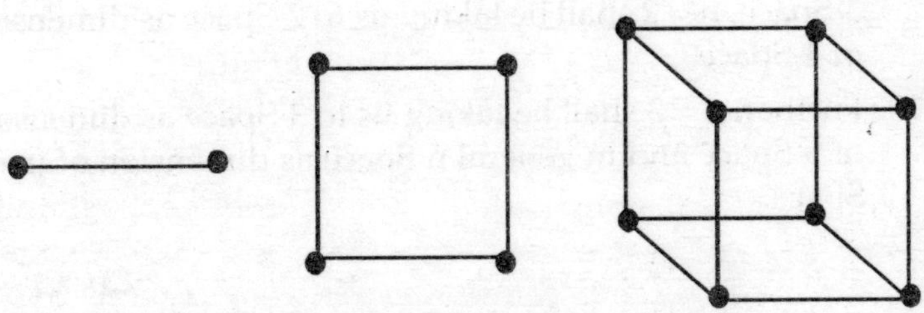

1. The arithmetic expression $(A+2)^n$ for n = 1 can be re-expressed as:

$$(A^1 + 2A^0)^1.$$

2. This expression gives us parallel geometric entities as $A^1$ as length of interval and as $2A^0$ as a pair of end points of a close interval.

3. The arithmetic expression $(A+2)^n$ for n = 2 can be re-expressed as:

$$(A^1 + 2A^0)^2 = (A^2 + 4A^1 + 4A^0).$$

4. This expression gives us parallel geometric entities as $A^2$ as area of a square and as $4A^1$ as four boundary lines of square and as $4A^0$ as four corner points of a square.

5. The arithmetic expression $(A+2)^n$ for n = 3 can be re-expressed as:

$$(A^1 + 2A^0)^3 = (A^3 + 6A^2 + 12A^1 + 8A^0).$$

6. This expression gives us parallel geometric entities as $A^3$ as volume of a cube and as $6A^2$ as six surface plates of a cube, as $12A^1$ as twelve edges of a cube and $8A^0$ as eight corner points of a cube.
7. The above property of need of adding two units to length A is a property which would help us to have comparative relationship of 1-space as dimension of (1+2) space.
8. In general, the above property of need of adding two units to the value A is a property which would help us to have comparative relation ship of n-Space as dimension of (n+2) Space.
9. In particular while n = 1 leads to 1-Space as dimension of 3-Space, n = 2 shall be taking us to 2-Space as dimension of 4-Space.
10. Further, n = 3 shall be taking us to 3-Space as dimension of 5-Space and in general n Space as dimension of (n+2) Space.

❑ ❑ ❑

# 57

# EXISTENCE OF HIGHER SPACES

1. We can view
   (*a*) interval (line) / 1-space as a track of a moving point,
   (*b*) surface (square) / 2-space as a track of a moving interval / line / 1-space, and,
   (*c*) solid space (cube) / 3-space as a track of a moving surface / square / 2-space.
2. From here we can pose a question to ourselves as to what would be the track of a moving solid / cube / 3-space?
3. The natural answer is 4-space / hypercube-4 / hyper solid domain.
4. We are having free movements of solids. Even when we move, we move as a solid block, so we always create 4-space for our free motion.
5. So our space in which we move freely is not a 3-space. It is 4-space and higher spaces all compactified.
6. With the motion of hypercube-4/4-space, we shall be having 5-space for us and in general with the motion of hypercube-n/n-space, we shall be having (n+1) space.
7. As such the existence of higher spaces is one concept for whose comprehension and chase one shall perfect one's intelligence beginning with the motions of point, line, square and cube.

8. Once one has complete comprehension of existence of 4-Space with hyper cube-4 as its representative regular body with domain boundary ratio as $A^4 : 8B^3$, one shall proceed further for comprehension of 5-Space and its representative regular body.
9. Once one has complete comprehension of existence of 5-Space with hyper cube-5 as its representative regular body with domain boundary ratio as $A^5 : 10B^4$, one shall proceed further for comprehension of 6-Space and its representative regular body.
10. Initially, for perfection of intelligence, one shall sequentially go on chasing dimensional spaces 1 to 7 and their representative regular bodies. The subsequent chase of higher spaces shall be taken up only after one has perfected one's intelligence in respect of 1 to 7 spaces.

❑ ❑ ❑

# 58

# GEOMETRIES OF 3-SPACE

One way to fix the dimensional space is to fix its representative regular body. Like that cube as representative regular body of three space helps us fix three space in terms of a cube. The dominant aspect of the dimensional body is its content lump. This in case of a cube turns out to be the three space content lump because of which is the solid portion/volume.

In terms of the domain/solid/content lump of a cube, we can have seven versions of the cube as:

**(1) C (0, 6):** A cube with all the six surface plates intact.

Alternatively, this cube also can be used as a cube with none of its surface plates being absent.

**(2) C (1, 5):** A cube with only five surface plates intact.

Alternatively, this cube also can be used as a cube with only one of its surface plates being absent.

**(3) C (2, 4):** A cube with only four surface plates intact.

Alternatively, this cube also can be used as a cube with only two of its surface plates being absent.

**(4) C (3,3):** A cube with only three surface plates intact.

Alternatively, this cube also can be used as a cube with only three of its surface plates being absent.

**(5) C (4, 2):** A cube with only two surface plates intact.

Alternatively, this cube also can be used as a cube with only four of its surface plates being absent.

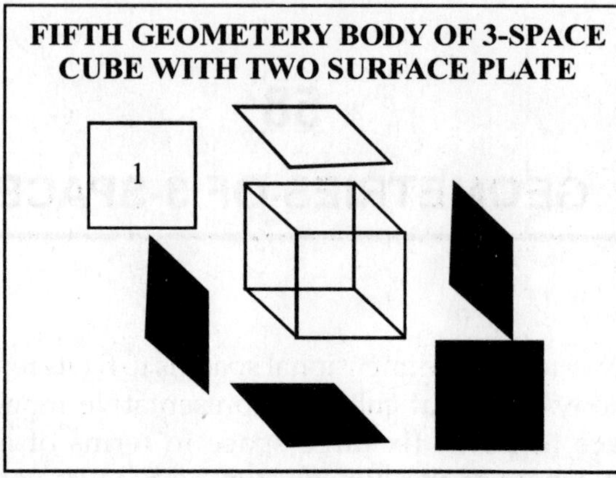


**(6) C (5, 1):** A cube with only one surface plate intact.

Alternatively, this cube also can be used as a cube with only five of its surface plates being absent.

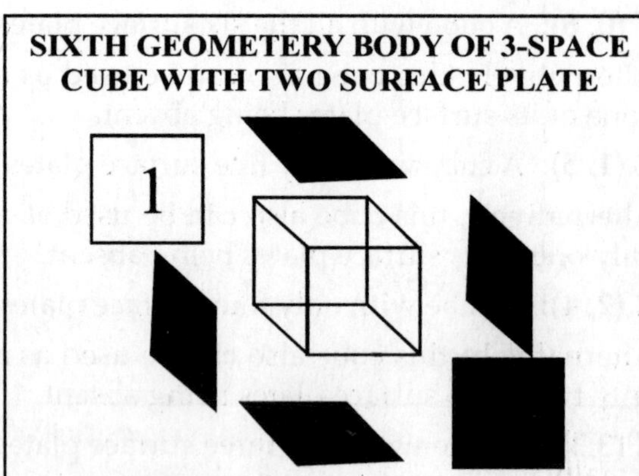


(7) C (6,0): A cube with all surface plates absent.

Alternatively, this cube also can be used as a cube with none of its surface plates being present.

SEVENTH GEOMETERY BODY OF 3-SPACE
CUBE WITH TWO SURFACE PLATE

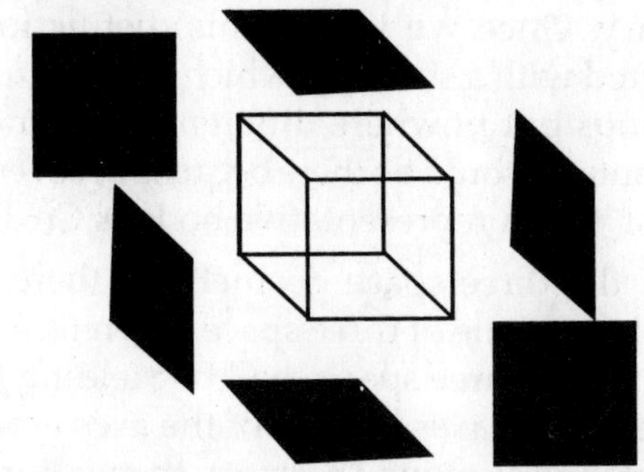

Above seven versions of the cube because of their distinct organization make them geometrically distinct and it is this distinction which makes their geometries distinct. As such we can say that three space admits seven distinct geometries.

We can tabulate as under:

| *Sr. No.* | *Geometry of three space* | *Representative body of three space* |
|---|---|---|
| 1 | G(0,3) | C(0,6) |
| 2 | G(1,3) | C(1,5) |
| 3 | G(2,3) | C(2,4) |
| 4 | G(3,3) | C(3,3) |
| 5 | G(4,3) | C(4,2) |
| 6 | G(5,3) | C(5,1) |
| 7 | G(6,3) | C(6,0) |

We can easily pictorially express seven versions of cube. Also we can formally define seven distinct geometries of three space.

***Note:***

[1] This classification of cube in terms of its surface plates (boundary components), would help us have insight

about our present day concept of continuity and differentiation. The continuity is there because of the domain and differentiation is there because of the boundary. Once we ignore this distinction, we stand confronted with a situation which takes us to everywhere continuous but nowhere differentiable functions which in the context would be there because of seventh geometry G(6,3) of which representative body is C(6,0).

[2] Each of the three space geometry is there within three dimensions frame of three space. Therefore, all the seven geometries of three space shall be yielding for us $3 \times 7 = 21$ dimensional axes. If each of the axes is accepted as of two parts (of di-monad format), then naturally we shall be having $21 \times 2 = 42$ components.

[3] Maheshwara Sutras are fourteen in number and these together coordinate 42 letters of Devanagari alphabet.

❑ ❑ ❑

# 59

# 2N + 1 GEOMETRIES FOR N SPACE

## 1. Interval as representative body of 1-Space

The domain boundary ratio of Interval accepts formulation $A^1 : 2B^0$.

As such Interval has three versions, namely, close interval, half-close interval, and open interval.

These three distinct versions of interval are representative bodies of 3 geometries of 1-space.

## 2. Square as representative body of 2-Space

The domain boundary ratio of Square accepts formulation $A^2 : 4B^1$.

As such Square has five versions, namely, Square with all boundary lines, with three boundary lines, with two boundary lines, with one boundary line, and with no boundary line.

These five distinct versions of Square are representative bodies of 5 geometries of 2-space.

## 3. Cube as representative body of 3-Space

The domain boundary ratio of Cube accepts formulation $A^3 : 6B^2$.

As such Cube has seven versions, namely, Cube with all the surface plates, and with five, four, three, two, one and no surface plate.

These seven distinct versions of Cube are representative bodies of 7 geometries of 3-space.

❑ ❑ ❑

# 60

# GEOMETRIES OF 4-SPACE VERSIONS OF HYPER CUBE-4

## 1. Hyper Cube-4 as representative body of 4-Space

The domain boundary ratio of Hyper cube-4 accepts formulation $A^4 : 8B^3$.

As such Hyper Cube-4 has nine versions, namely, Hyper Cube-4 with all the eight solid boundary components, and with seven, six, five, four, three, two, one and no solid boundary component.

These nine versions of Hyper Cube-4 are representative bodies of 9 geometries of 4-space.

### FIRST GEOMETRY

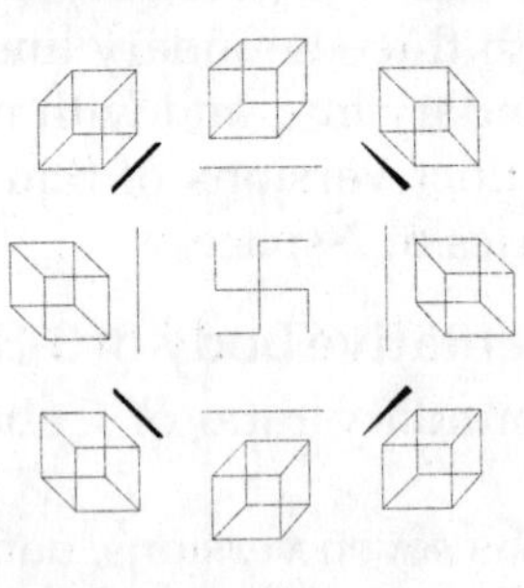

## SECOND GEOMETRY

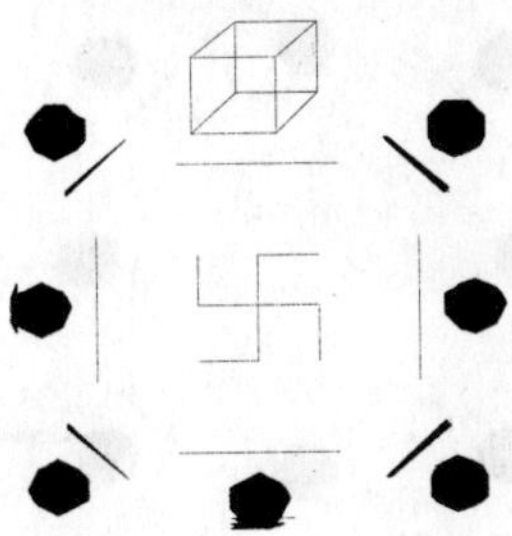

## THIRD GEOMETRY

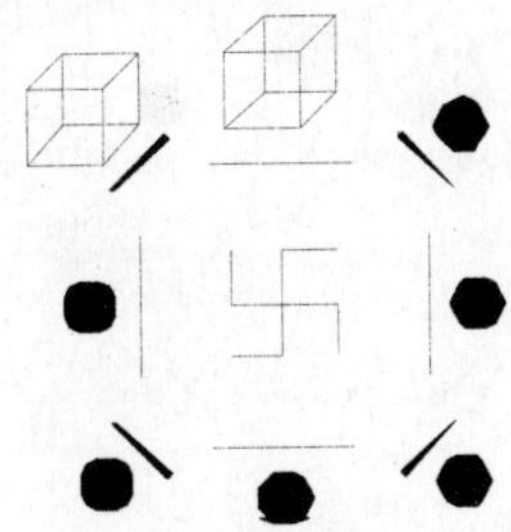

## FOURTH GEOMETRY

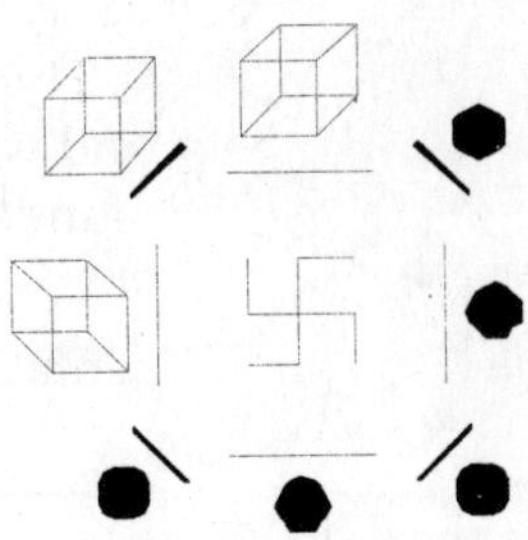

## FIFTH GEOMETRY

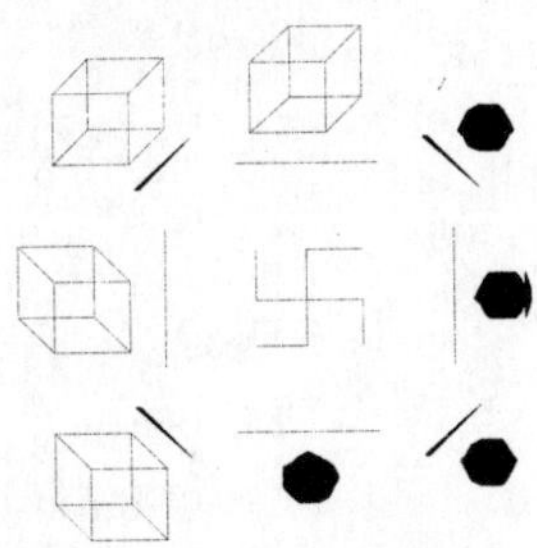

## SIXTH GEOMETRY

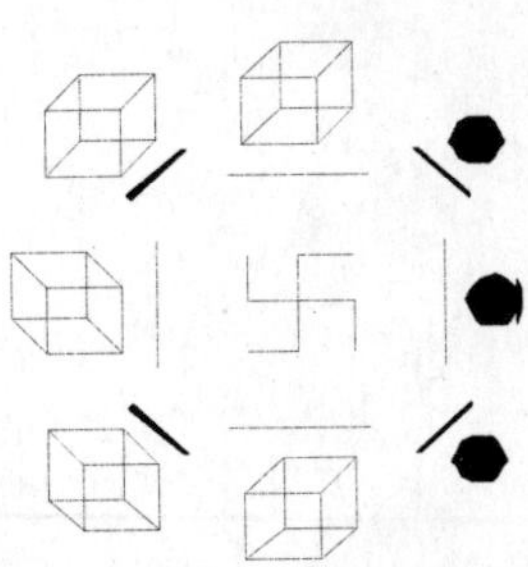

## SEVENTH GEOMETRY

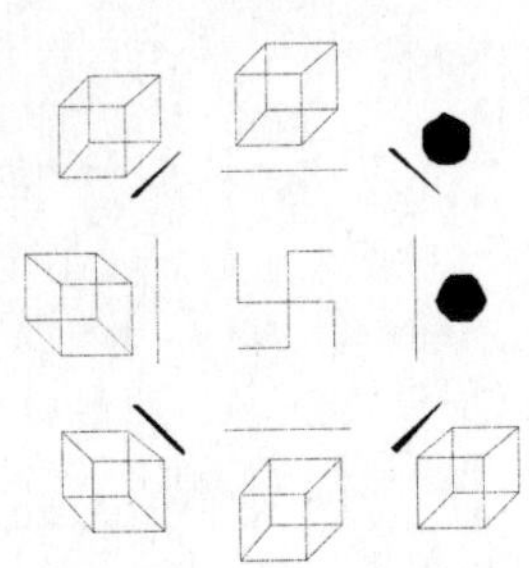

| EIGHTH GEOMETRY | NINTH GEOMETRY |
|---|---|
| 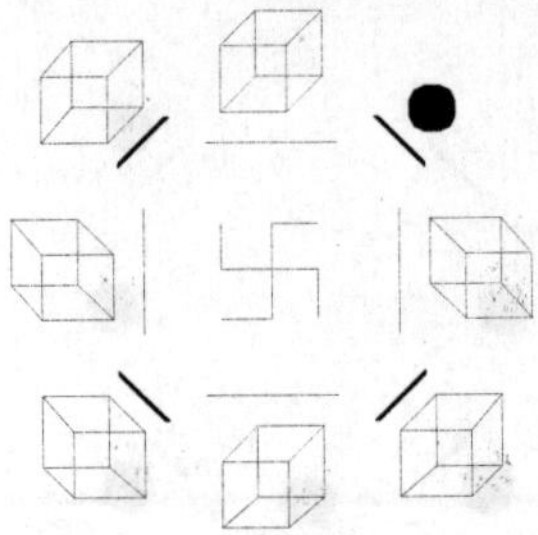 | 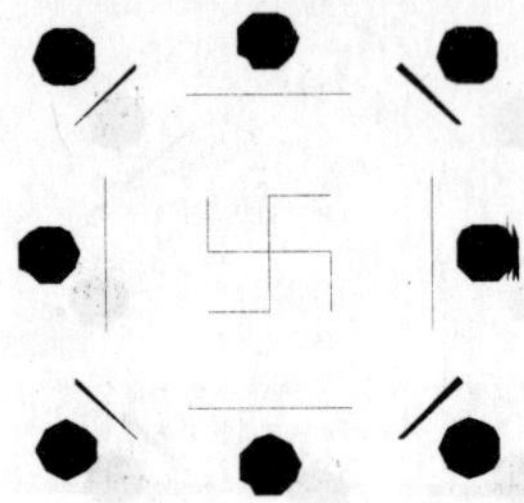 |

# 61

# GEOMETRIES OF 4-SPACE VERSIONS OF HYPER CUBE-5

## 1. Hyper Cube-5 as representative body of 5-Space

The domain boundary ratio of Hyper cube-5 accepts formulation $A^5 : 10B^4$.

As such Hyper Cube-5 has eleven versions, namely, Hyper Cube-5 with all the ten hyper solid-4 boundary components, and with nine, eight, seven, six, five, four, three, two, one and no hyper solid-4 boundary component.

These eleven versions of Hyper Cube-5 are representative bodies of 11 geometries of 5-space.

### FIRST GEOMETRY

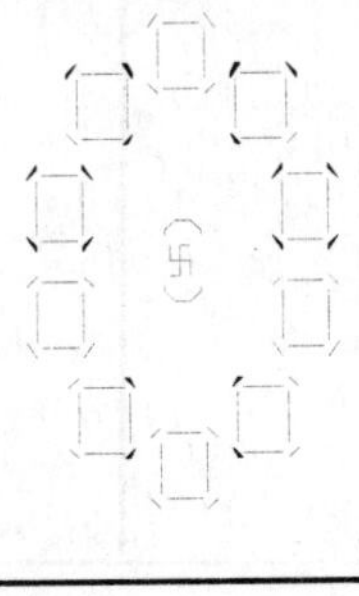

## SECOND GEOMETRY

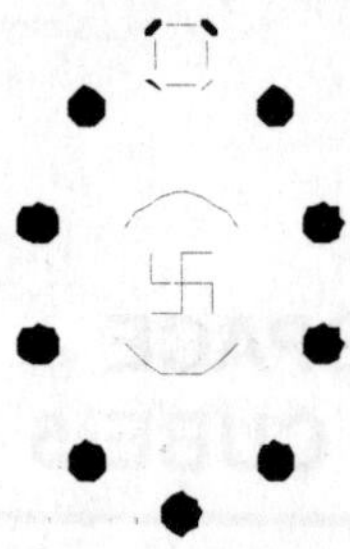

## THIRD GEOMETRY

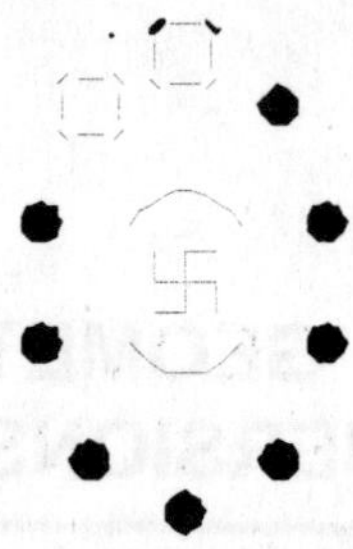

## FOURTH GEOMETRY

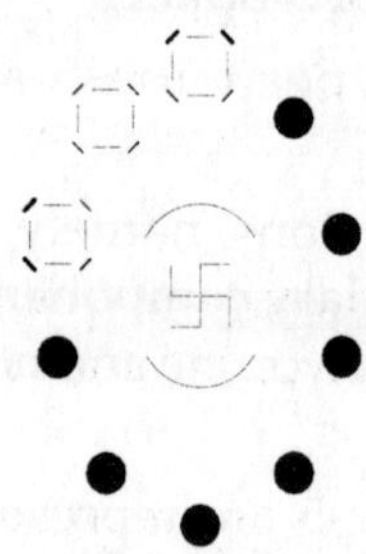

## FIFTH GEOMETRY

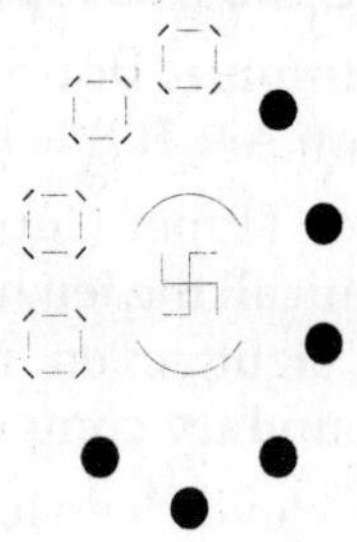

## SIXTH GEOMETRY

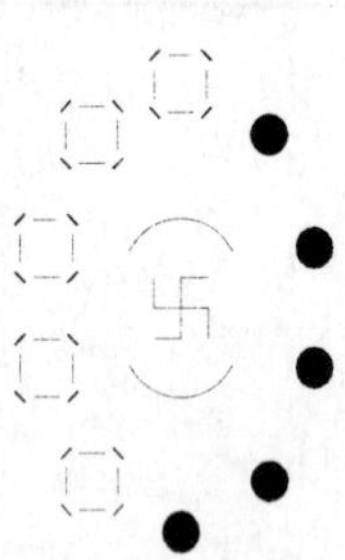

## SEVENTH GEOMETRY

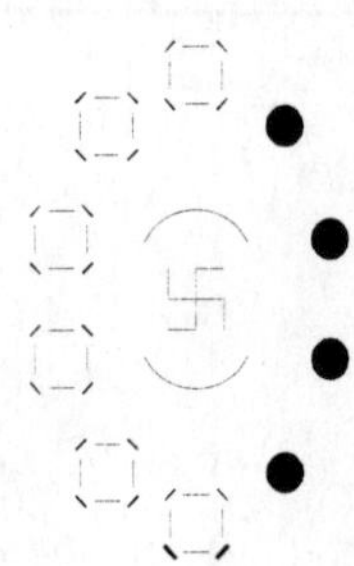

## EIGHTH GEOMETRY

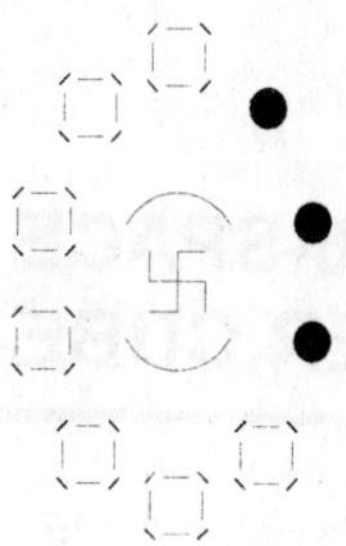

## NINTH GEOMETRY

## TENTH GEOMETRY

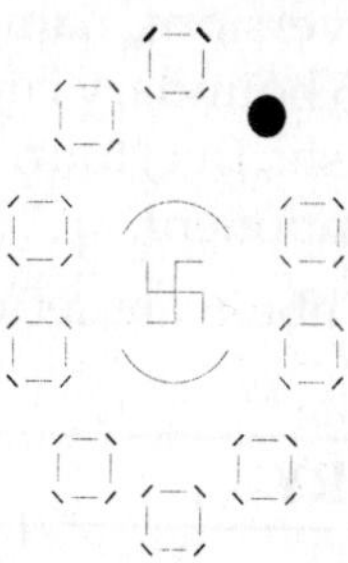

## ELEVENTH GEOMETRY

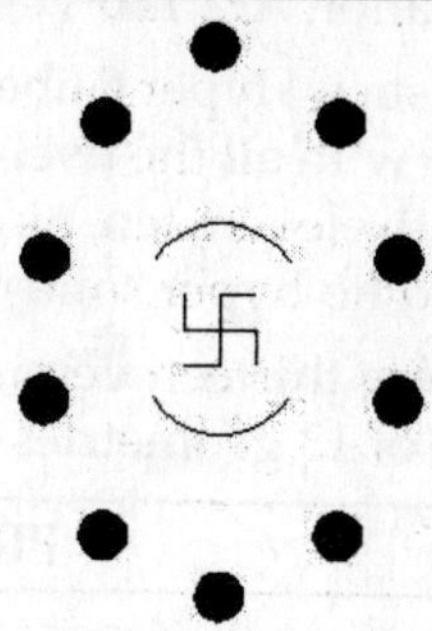

# 62

# GEOMETRIES OF 6-SPACE VERSIONS OF HYPER CUBE-6

## 1. Hyper Cube-6 as representative body of 6-Space

The domain boundary ratio of Hyper cube-6 accepts formulation $A^6 : 12B^5$.

As such Hyper Cube-6 has thirteen versions, namely, Hyper Cube-6 with all the twelve hyper solid-5 boundary components, and with eleven, ten, nine, eight, seven, six, five, four, three, two, one and no hyper solid-5 boundary component.

These thirteen versions of Hyper Cube-6 are representative bodies of 13 geometries of 6-space.

### FIRST GEOMETRY

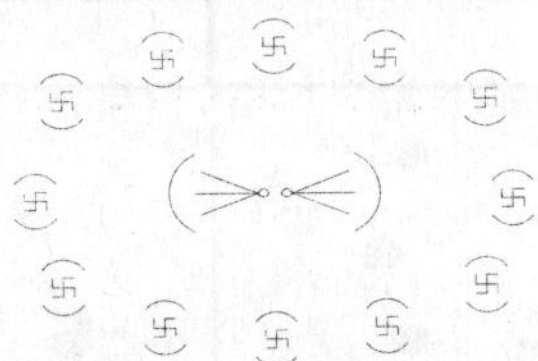

### SECOND AND THIRD GEOMETRY

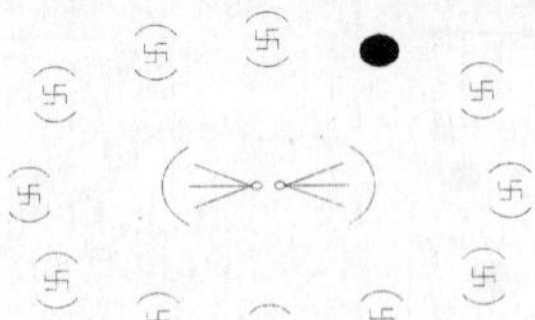

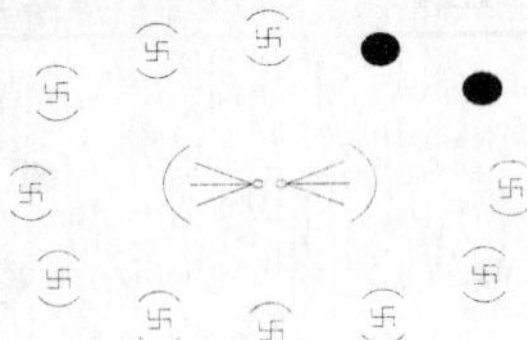

## FOURTH GEOMETRY

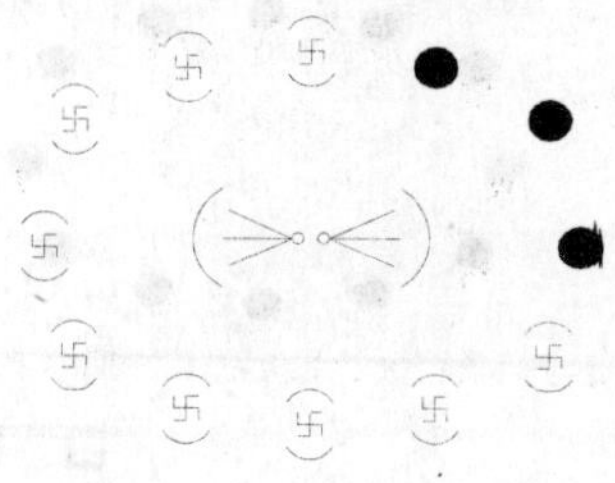

## FIFTH GEOMETRY

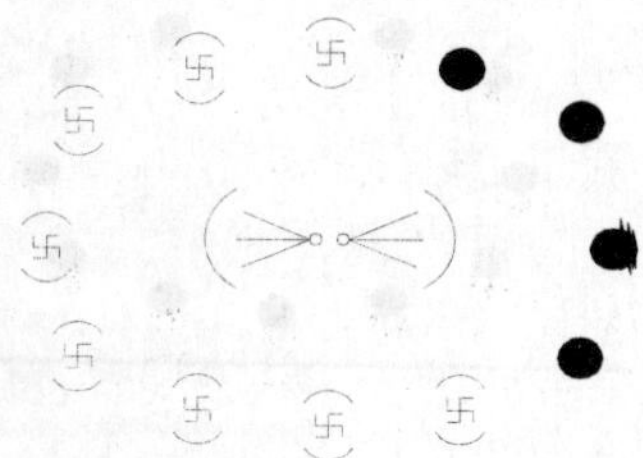

## SIXTH GEOMETRY

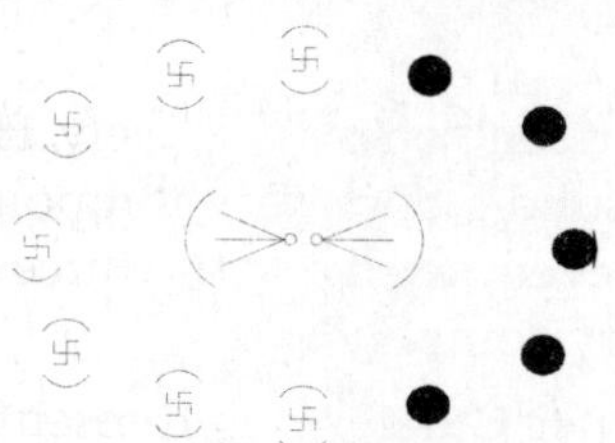

## SEVENTH GEOMETRY

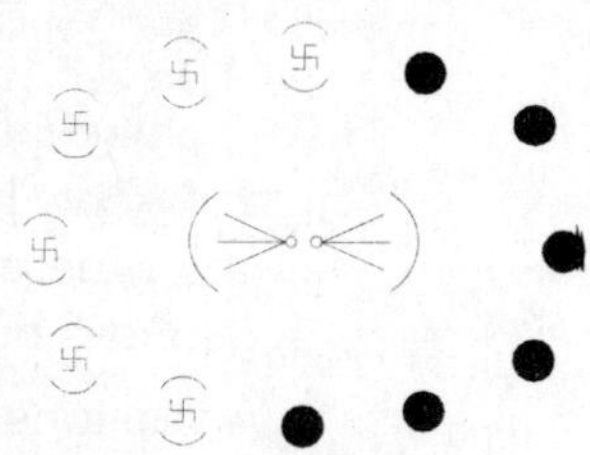

## EIGHTH GEOMETRY

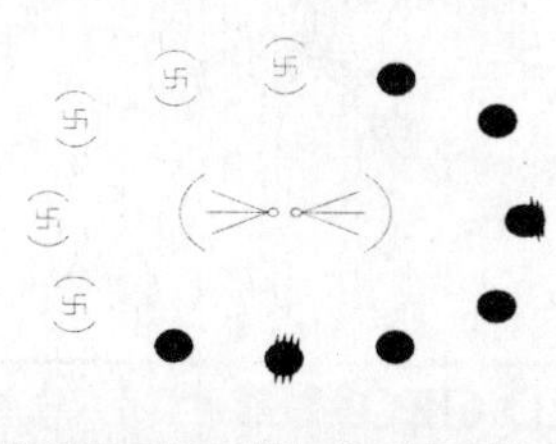

## NINTH GEOMETRY

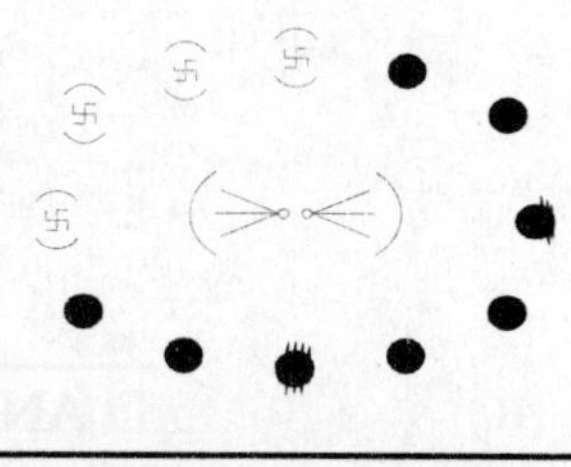

## TENTH GEOMETRY

## ELEVENTH GEOMETRY

| TWELFTH GEOMETRY | THIRTEENTH GEOMETRY |
| --- | --- |
| 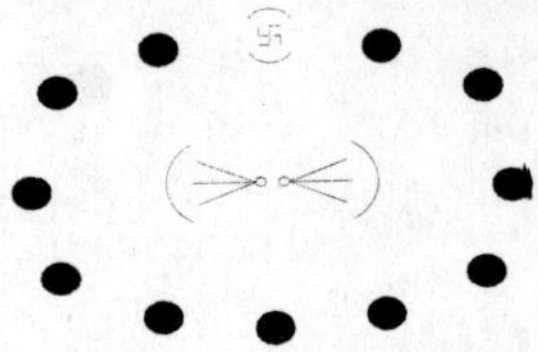 | 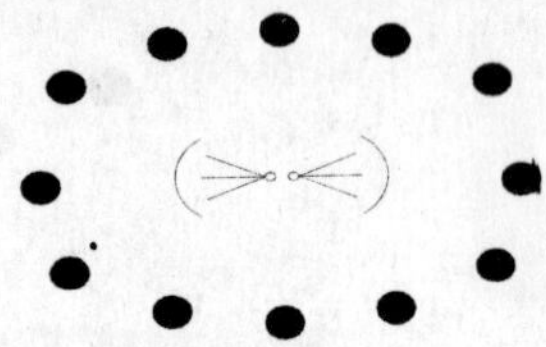 |

# 63

# MANIFESTED CREATIONS

1. Creator space (4-space) provides four fold manifestations format.
2. Along the four fold manifestation format manifest hyper cubes.
3. The four folds of manifested bodies of dimensional spaces, designated and known as corresponding hyper cubes are (*i*) Dimension (*ii*) Boundary fold (*iii*) Domain fold and (*iv*) Origin Fold.
4. These four folds are set ups of four consecutive dimensional spaces contents.
5. Cube is the manifested body of four folds namely 1-space content being its dimension fold, 2-space content being its boundary fold, 3-space content being domain fold and 4-space content being the origin fold.
6. Cube as such as a manifested body is hyper cube-3.
7. The four folds of cube as hyper cube-3 are expressible as quadruple of artifices (1, 2, 3, 4).
8. Likewise hyper cube 4, the representative regular body of 4-space is expressible as a quadruple artifices (2, 3, 4, 5) as here 2-space plays the role of dimension, 3-space as boundary fold, 4-space as domain fold and 5-space as origin fold.
9. Further the quadruples (3, 4, 5, 6) and (4, 5, 6, 7) respectively represent hyper cubes 5 and 6 respectively.

❑ ❑ ❑

# 64

# OVERVIEW OF THE APPROACH TO THE DISCIPLINE OF GEOMETRY

## 1. Introductory

1. Vedic Systems approach is multi-dimensional.
2. Vedas avail real spaces as Organization formats.
3. Parallel to Numerals of the place value system emerge corresponding real-dimensional spaces.
4. These real dimensional spaces are of sequential dimensional orders.
5. Our well known three dimensional bodies are of linear dimensional set-ups, that is, here one space plays the rule of dimension.
6. The Geometry which restricts itself uptil three space, that is, uptil solids, is excepted as a linear order Geometry. It is designated and is known as "Rekhaganit"/ The Mathematics of line.
7. Vedic Geometry, as such, is not restricted uptil linear dimensional order. Beyond that special dimensional order of four space, solid order of five space, and hyper solid orders from six space onwards as well are the real interest of Vedic Geometry.
8. However as, in the present first volume, the chase is restricted only uptil the linear order as such the focused exposure is entered upon solids, and of these as well, only upon the lines, surfaces and solids with the running

formats of interval, square and cube as well as of Point, line, circle and sphere.

## 2. Artifices of Numbers and Dimensional Frames

1. Vedic Processing systems are of two fold features, designated and known as Sankhyanishta and Yoganishtha.
2. Sankhya nishtha presume the existence of Geometric formats and avails values of numbers.
3. Yoga nishtha presumes the existence of numbers and avails the values of the Geometric formats.
4. Of it the specific applied values of numbers and Geometric formats being availed as of the artifices of number and dimensional spaces.
5. The artifices of numbers are chased presuming the existence of dimensional spaces, and on the other hand the dimensional frames are chased presuming the existence of artifices of numbers.
6. Standing restricted for the chase only uptil the linear order three space, the dimensional frame made use of three dimensions as three axes of three lines format.
7. Corresponding to it the artifices of first three numbers namely (1, 2, 3)/(1, 1 + 1, 1 + 1 + 1)/(1, 1 × 1, 1 × 1, 1)/ $(1^1,1^2,1^3)$/[(1), (1, 1), (1, 1, 1)]

## 3. Ganita Sutras Approach

8. While going for arithmetic chase, the Geometric formats are presumed.
9. It would mean that, it is being presumed that Ganita Sutras rules are as per the features of Geometric formats being Sequentially availed from first Sutra to the last Sutra, as well as from first Upsutra till the last Upsutra.
10. With availability of the sequential Geometric formats, the values of arithmetic strating with sequential features of artifices of numbers are worked out.

11. It would mean that arithmetic is being done on the Geometric formats
12. On the other hand while the Geometry, add in particular Rekhaganit/ Mathematics of line is being worked, it is being presumed that values of numbers, and in particular the features of artificies of numbers are available.
13. It is in this background that the working rule of Ganita Sutra-1 "One more than before" shall be sequentially unfolding "points as zero degree bodies, lines as first degree bodies, planes as two degree bodies, solids as four degree bodies, hypersoldis four as four degree bodies and so on."
14. Ganita Upsutra-1 with its rule of Symmetry and Proportion chase by following the forms as these are friend shall be sequentially taking us from lines to parallel lines to angles, to triangles, to quadrilaterals, to pentagons, and so on to circles, to spheres, to hyperspheres and so on.
15. Ganita Sutra-2 with the working rule of "all from Nine and last from ten", making available ten place value system for Organizations of artifices of numbers and for values of numbers, shall be correspondingly leading to the chase of Geometric formats through angles.

Here it may be relevant to mention for pointed attention.

## Total Angle Value in Different Place Values

Parallel to ten place value system, total angle at a point in the surface may be had in terms of different place value systems. On six place value system as there are five numerals as such the value of total angle for all the four quarters shall be $4 \times (5 \times 6) = 120$. However as in six place value system "6" ="10", as such it shall be $4 \times (50)$ = "320" in six place value system. Like that the value may be computed in different place value system.

The sequential order of Ganita Sutras as along artifices 1 to 16, and corresponding to it the parallel Geometric formats thereof, shall be sequentially leading as Ganita Sutra-1 organizing features of line, sequential order of line, linear dimensional order. Ganita

Sutra-2 shall be leading to Surfaces, special orders, angular coverage parallel to different place value system. Ganita Sutra-3 leading to solid order with transition from horizontal plane to vertical plane. Ganita Sutra-4 shall be taking a step ahead to a four space with reflection operation and working with half dimensions and so on. A step ahead would be a transcends from avyakta to aviakto-avyakat, that is from special order for space to solid order five space, where the zero space and zero space content shall be having their roles to play, which symmetrically is to be availed for six space values to which Ganita Sutra-6 leads to. Beyond that, that is from Sutra-7 onwards is to be self referral feature of plus one space and negative one space are two simultaneously come into play and with it the chase is to really enter the real spaces features of hyper dimensional orders, which chase is being left to be taken of in the other volumes.

❑ ❑ ❑

# 65

# HOW TO START LEARNING AND TEACHING OF THE DISCIPLINE OF GEOMETERY TO THE YOUNG MINDS

## 1. Introductory

1. Vedic Geometry and its systems, as has been indicating in lesson 1 under overview of the approach of the discipline, are really such for which a very gentle approach is expected for young minds from the teachers
2. However, simultaneously, it is also important as that the discipline can not be afforded to be permitted to get closed in a harden linear order as it may result into such mental blocks which after school education would be difficult to transcend through.
3. Time and again, exposure is to be given to the young minds to simultaneously change the setups of closed interval, square and cube as well as of circle and sphere to impress upon as that all these are governed by the single formulation for their boundaries and domains ratios, in terms of which the smooth transition can be had to the representative regular bodies of four and higher spaces, as well as to the existence and blissful features of those spaces.
4. No doubt, uptil the middle school the central focus is to be upon the setup of "cube" as well as of sphere, but the same or the members of the sequence of representative reguar bodies of whole range of dimensional spaces and that the world and discipline of Geometry is not limited

uptil the mathematics of line/linear order/linear order three space like the numbers sequence is not exhausted and stopped at artifice 3/whole number 3.

## 2. Cube

5. Cube is to be the thumb rule of Mathematics uptil linear order.
6. Cube may be taken as the text book of three space features.
7. As such, Cube is to be the best friend of the students of this phase and stage of learning.
8. It becomes the solemn responsibility of the teachers to very gently expose the young minds to the beautiful features and blissful values of the setup of the cube.
9. Here are being enlisted some of the features of the cube.

❑ ❑ ❑

# 66

# BASIC FEATURES AND MEASURES OF THE GEOMETRIC FIGURES AND BODIES

## Definitions

1. Point: "Point is the Geometric entity which is divide of length, breadth and even height."

   Its measure as such is taken as "Zero"

2. **Line:** "Line: has length but has no breadth or height.As such, it has expression for its value as first degree values of numbers ($n^1$)
3. **Surface:** "Surface" has length as well as breadth but has no height. As such it has expression for its value as product of two single degree numbers ($l \times b$)
4. **Parallel lines:** "Parallel lines" do not Intersect or meet each other. Parallel lines are also taken by definition as the lines which meet at infinity.
5. **Transversal lines:** "Transversal lines" is the line which intersects the given Parallel lines.
6. **Parallel Transversal lines:** "Parallel Transversal lines" intersect given Parallel lines at equal distances.
7. **Intersecting lines:** Pair of Intersecting lines make four angles at the intersecting points of which the opposite angles are equal.
8. **Total angle at a point:** "Total angle at a point" may be defined as Parallel to the circumference of a circle with given point as its centre. As per the Ganita Sutra-2 rule of ten place value system with nine numerals shall be

measuring at the given point as centre/middle point of diameter/given interval permitting approach in four parts, of which two parts being from the centre to the end points and in reverse orientation as well in two parts from end points to the centre, and thereby value to be 4(4 × (9 × 10)) = 360 degree. Conventionally it is being taken as four Right angles corresponding to four quarters of the circumference. It is also taken as two pairs of Right angles Parallel to the pair of orientation approach from middle to the end points of the interval/diameter and as reverse there of from end point of intervals to the middle. With it comes the acceptance for the total angle at a point in the surface being 2∏.

9. **Total angle value in different place values:** Parallel to ten place value system, total angle at a point in the surface may be had in terms of different place value systems. On six place value system as there are five numerals as such the value of total angle for all the four quarters shall be 4 × (5 × 6) = 120. However as in six place value system "6" = "10", as such it shall be 4 × (50) = "320" in six place value system. Like that the value may be computed in different place value system.

10. **Types of Triangles:** On the basis of the sides, we shall be having three types of triangles, namely

    (*a*) **Scalene triangle:** A triangle in which all the sides are of different lengths.

    (*b*) **Isosceles triangle:** A triangle whose two sides are equal is called an Isosceles triangle

    (*c*) **Equilateral triangle:** A triangle whose all sides are equal is an Equilateral triangle.

11. **On the basis of angles:**

    (*a*) **Acute traingle:** A triangle whose all the three angles are less than Right angle is an Acute triangle.

    (*b*) **Right angled triangle:** A triangle whose one angle is Right angle is Right angled triangle.

(*c*) Obtuse angled triangle: A triangle whose one angle is obtuse is Obtuse angled triangle.

12. **Sum of the angles of triangle:** Sum of the angles of a triangle is two Right angles.
13. **Exterior angles of triangle:** Sum of the Exterior angles of triangle is four Right angles.
14. **Congruent lines:** Two lines segments are Congruent when they are of equal length.

| Sr. No. | Quadrilateral | Lengths of sides | Degrees of Angles | Parallel Nature of sides |
|---|---|---|---|---|
| 01 | Quadrilateral | may or may not be equal | may or may not be equal | may or may not be parallel |
| 02 | Square | All sides are equal | All angles are equal | Opposite sides are parralel |
| 03 | Rhombus | ----do----- | Opposite angles are equal | -------do------- - |
| 04 | Rectangle | Opposite sides are equal | All angles are equal | -------do----- |
| 05 | Trapezium | May or may not be equal | May or may not be equal | One pair of opposite sides are parallel |
| 06 | Parallelogram | Opposite sides are equal | Opposite angles are equal | Opposite sides are parralel |
| 07 | Kite | Two pairs of adjacent sides are equal | Opposite angles are equal | Opposite sides are parallel |

15. **Congruent angles:** Two angles are Congruent when they are of the same degree.
16. **Congruent triangles:** Two triangles are Congruent when their three sides are correspondingly equal and their three angles are also correspondingly equal.
17. **Two sides and One angle basis:** Two triangles with their two corresponding sides being Congruent and their one angle as well being Congruent, it would make two triangles being Congruent.
18. **One side and pair of angles at its end points:** Two triangles when one of their sides is Congruent and two angles at its end points are also Congruent, then the triangle themselves are Congruent.
19. **Quadrilaterals:** Quadrilaterals the following table and lists different types of Quadrilaterals with features.
20. **Pyramid:** It is a solid figure. It has One Vertix outside the plane of the base. Its faces connecting the sides of the rectilinear base are triangles.
21. **Tetrahedron:** "Tetrahedron" is a Pyramid with triangular base.
22. **Right Pyramid:** "Right Pyramid" is the Tetrahedron with base as an Equilateral triangle. Lateral surface of right Pyramid = 3/2 × (side of the base triangle) × (slant height of the pyramid)

    Volume of the Right Pyramid is = √3/12 × (square of the side of the base triangle) × height of the pyramid.
23. **Regular Tetrahedron:** A Regular Tetrahedron is a Tetrahedron with equal edges.
24. **Cube:** "Cube" is a solid with equal length, breadth and height.
25. **Cuboids:** "Cuboid" is a solid with its all six surfaces being rectangular.
26. **Right Circular Cylinder:** "Right Circular Cylinder" is a solid with equal circular base and top and is bounded by

a curved surface. The surface area of the curved surface is equal to (circumference of the base circle) × (height of the cylinder) and the volume of the Right circular cylinder is equal to (area of the base circle) × height of the right circular cylinder.

27. **Right Circular Cone:** "Right Circular Cone" is a solid with Circular base and a single vertex outside the base and being bound by a curved surface. Its volume is equal to 1/3 × (area of the circular base) × (height of the cone).
28. **Sphere:** "Sphere" is a solid bounded by a closed surface at equal distance from the centre. Its surface area is equal to four times the area of the central circle and its volume is equal to 4/3 × (area of the central circle) × (radius of the sphere).
29. **Central Circle of the Sphere:** "Central Circle of the Sphere" is the biggest circle which can be drawn within the sphere at centre of the sphere.
30. **Secant of the Circle:** "Secant of the Circle" is a line which cuts the circle at two points
31. **Chord of a Circle:** "Chord of a Circle" is the part of the Secant of the Circle between the two points at which the Secant cuts the circle.
32. **Arc of the Circle:** "Arc of the Circle" is part of the circumference of the circle between the two points of the Secant of the circle. The chord and the Arc of the circle are interconnected at two points of the secant of which the chord is the part.
33. **Cyclic Quadrilateral:** "Cyclic Quadrilateral" is the quadrilateral whose all the four vertices are on the circle.

❑ ❑ ❑

# 67

# ACQUINT ONESELF WITH THE BASIC GEOMETRIC FIGURES AND BODIES

## 1. Introductory

1. Point, line, Surface and solids are four different types of Geometric Entities.
2. Of these, points are divide of length, breadth as well as of height.
3. Line is having only length. It is divide of breadth as well as of height.
4. Surface has area. As such it has length and breadth but is divide of height.
5. Solids have length, breadth as well as height.

## 2. Geometric Figures

6. It be taken that figures are made of simultaneous values of length and breadth.
7. As such it would emerge that figures are formed in surfaces.
8. The basic constituents of figures are (*i*) Points (*ii*) Lines (*iii*) Angles and (*iv*) Surface Area of the figure.
9. There are two types of figures, known as (*i*) Open figures and (*ii*) Closed figures
10. Open figures do not encircle surface area.
11. The closed figures encircle surface area.
12. Common Open figure is that of an angle.

13. The common closed figures are (*i*) Triangle (*ii*) Quadrilateral (*iii*) Pentagon (*iv*) Hexagon (*v*) Circle (*vi*) Closed Semi-Circle

## 3. Area of Closed Figures

14. Square and Right Angle Triangle are two specific Closed figures with the help of which the area of the Closed figures.
15. Area of Square is obtained as Square as Square of the Side of the Square.
16. The area of Right Angle Triangle is obtained as Half of the Area of the Rectangle.
17. The area of Right-Angle Triangle of Equal Sides is obtained as half of the Area of the Square.

## 4. Different Types of Quadrilaterals

18. Quadrilateral is a figure which has four sides.
19. As Quadrilateral has four sides, as such it also has four angles.
20. When all the four sides are equal and all the four angles as well are equal, the Quadrilateral is designated as a Square.
21. When only the four angles are equal, the Quadrilateral is designated as a Rectangle.
22. Rectangle becomes a Square when all the four sides are also equal.
23. Quadrilateral with all the sides equal and opposite sides parallel is designated as a Rhombus.
24. When the Opposite sides of the Quadrilateral are parallel, it is designated as a Parallelogram.
25. When One pair of side of the Quadrilateral are parallel and other pair of sides are not parallel then the quadrilateral is designated as a trapezium.
26. When the two pairs of sides of the Quadrilateral are parallel and two pairs of adjacent sides are equal then the Quadrilateral is designated as a Kite.

## 5. Distinguishing Features of Different Types of Quadrilaters

27. From the above designations of different types of Quadrilaterals, it may be noted as that these distinguishing features of these figures are there because of the length of the sides, the degrees of the angles and the feature of sides be Parallel or not.

28. On the basis of the distinguish feature of above different types of Quadrilaterals may be tabulated as follows:

| Sr. No. | Quadrilateral | Lengths of sides | Degrees of Angles | Parallel Nature of sides |
|---|---|---|---|---|
| 01 | Quadrilateral | may or may not be equal | may or may not be equal | may or may not be parallel |
| 02 | Square | All sides are equal | All angles are equal | Opposite sides are parralel |
| 03 | Rhombus | -----do----- | Opposite angles are equal | --------do-------- |
| 04 | Rectangle | Opposite sides are equal | All angles are equal | -------do----- |
| 05 | Trapezium | May or may not be equal | May or may not be equal | One pair of opposite sides are parallel |
| 06 | Parallelogram | Opposite sides are equal | Opposite angles are equal | Opposite sides are parralel |
| 07 | Kite | Two pairs of adjacent sides are equal | Opposite angles are equal | Opposite sides are parallel |

## 6. Surfaces Areas and Volumes of Solids

29. The surface of a solid may be measured with the help of suitable cuttings of paper/cloth in terms of which the surface of the solid may be covered. The surface area of the solid as such shall be equal to Surface area of the cloth/paper needed to cover the surface of the solid.

30. The volume of the solid may be measured with the help of the volume of the water which would completely fill the solid portion of the solid with in its surface area.

31. Cube is the representative solid.

32. Volume of the Cube is equal to the Cube of the side of the Cube.

    Cube has six surfaces each equal to the Square of the Side of the Cube and as such the total surface area of the Cube would be six times the single surface of the Cube.

❑ ❑ ❑

# 68

# APPLICATIONS OF GEOMETRIC FORMATS

## 1. Introductory

1. Applications of Geometry means making use of the Geometric formats.
2. Of these, to start with, the basic formats are to be availed.
3. Amongst these basic formats, is the format of a point.
4. The format of point, because of it being divide of length, breadth and even height.
5. As such, point is to be taken as of the features of "0" vis-a-vis the measures of length, area and volume.
6. As 0+0=0 and also as 0´0=0, as such there may be a number of points superimposed upon each other without changing the "Geometric set up/format of the point."
7. With it, it can be taken that a number of lines, in fact, infinitely many lines may be drawn from a single point.
8. This feature of "Point", is very useful for making construction of Geometric figures which have corner points, joints, knots and like that.
9. As there may be more than one lines meeting at a point, as such the format of Intersecting lines shall be making a very valuable format.
10. The basic feature of this format of Intersecting lines, to the specific, a pair of lines meeting at a point shall be

constituting a Geometric set up, known as angle would become a handy Geometric format.

11. With the help of an angle, an additional Geometric tool would be available to cover a surface.
12. It would help divide the closed surface in terms of angular boundary.
13. This angular boundary, at its limit shall be providing us another Geometric set up, known as the circle.
14. From angle to circle with intermediate angular phasing/ angularly phased boundary of as many phases, as may be required, shall be providing us from triangular to Quadrilaterals to Polygons of all number of sides and angles.
15. It is this chain starting with a point, to line, to angle, to triangle and to polygons shall be making our Geometric tool box to be very reach.
16. Of these point, line, angle, triangle, and quadrilateral, in particular are going to be the most valuable formats with us for constructing different Geometric tools for our applied purposes.

## 2. Help of Ganita Sutras Format

17. For construction of different Geometric tools of different Geometric formats, the formats and values of Ganita Sutras may be availed.
18. The sequential order of Ganita Sutras may be worked out simultaneously as per Sankhya Nishtha and Yoga Nishtha processes and approaches.
19. For our solar Universe of six steps range from "Earth to Sun", Ganita Sutras 1 to 6 range is to provide the mathematics science and technology for the needed Geometric formats for the appropriate Geometric tools.
20. Six steps long range from "Earth to Sun" is of "Earth, Water, Fire, Air, Space, Sun".
21. These sequentially or of the orders "Linear, Spatial, Solid and Hyper solids 4,5,6"

22. As such "one space, tow space, three space, four space, five space, six space" formats for dimensional orders are to be avail.
23. This would mean that "interval, square, cube, hypercubes 4, 5 and 6" are to be availed.
24. Correspondingly the formats and values of Ganita Sutras-1, 2, 3, 4, 5, 6 are respectively to be avail.
25. Parallel to it the artifices "1, 2, 3, 4, 5, 6" are to be availed
26. As for 2, 1 and 2 both would be available, and for 3,1,2 and 3 would be simultaneously available and like that for subsequently artifices as well, as such the formats, values and features of Ganita Sutras are to be sequentially availed as step 1, the Ganita Sutra-1, as step 2, the Ganita Sutras-1 and 2, as step 3, the Ganita Sutras-1, 2 and 3 and so on.
27. With it, for Earth element, being the first element, Ganita Sutra 1 shall be helping us to provide us the linear order format, values and features.
28. The Water element, being the second element, for it Ganita Sutra-2, in particular, and together with Ganita Sutra-1, shall be providing format, values and features for its chase as flow surfaces as of streams.
29. A step ahead, for fire element, Ganita Sutra-3 format values and features, in particular and along with that of Ganita Sutra-1 and 2 shall be providing us the format, values and features for the construction of the needed Geometric Instruments for chase of this element/fire element.
30. A step ahead, like wise is to be the help of Ganita Sutras-4,5 and 6 which is to be availed for the total applied values of the pure values of the vedic mathematics, science and technology formats, values and features of Ganita Sutras-1 to 6 in particular.

    The present approach of vectors, tensors, matrices, topology, measures and like that are the particular feature of the Geometric formats of real spaces, which to be taken up and to be dealt with in the subsequent volumes.

❑ ❑ ❑

# 69

# OVERVIEW OF GANITA SUTRAS DOMAIN

1. One way to approach Ganita Sutras domain is to proceed approaching this domain in the sequence and order of Ganita Sutras themselves.

## 1. Ganita Sutra-1

**SUTRA - I**

एकाधिकेन पूर्वेण

| 01 | 02 | 03 | 04 | 05 | 06 | 07 | 08 | 09 | 10 |
|---|---|---|---|---|---|---|---|---|---|
| ए | क् | आ | ध् | इ | क् | ए | न् | अ | प् |
| 11 | 12 | 13 | 14 | 15 | 16 | | | | |
| ऊ | र् | व् | ए | ण् | अ | | | | |

| Total Letters | Vowels | Nasels | Consonents |
|---|---|---|---|
| 16 | 8 | | 8 |

## 2. Initial Steps Outline

(*i*) Read the text of the Sutra.

(*ii*) Pronounce the text Loudly.

(*iii*) Chase the text for its working rule

(*iv*) Think, meditate and transcend through the text of the Sutra to have complete insight and for fully imbibing its values to perfect one's skills for applications of the working rule of the Sutra.

(*v*) And to have the overview as follows

2. The processing steps, as per the working rule of Ganita Sutra-1 shall be in the sequence and order of 'one more than the previous'.
3. This unfolding as per the established processing process of Sankhiya Nishtha and Yoga Nishtha shall be, firstly to go for the unfolding of artifices of numbers presuming the existence of the dimensional domains.
4. The next step shall be to go for unfolding of dimensional frames presuming the existence of artifices of numbers.
5. Once both artifices of numbers and dimensional frames, as such shall be available, the processing is to simultaneously go for 'relation inter see of artifices of numbers and of dimensional frames.

### 3. Ganita Upsutra-1

**UPSUTRA - I**

आनुरूप्ये

| 01 | 02 | 03 | 04 | 05 | 06 | 07 | 08 | 09 | 10 |
|---|---|---|---|---|---|---|---|---|---|
| आ | न् | उ | र् | ऊ | प् | य् | ए | ण् | अ |
| Total Letters | | | Vowels | | Nasels | | Consonents | | |
| | 10 | | 5 | | - | | 5 | | |

## 4. Initial Steps Outline

(*i*) Read the text of the Upsutra.

(*ii*) Pronounce the text Loudly.

(*iii*) Chase the text for its working rule

(*iv*) Think, meditate and transcend through the text of the Upsutra to have complete insight and for fully imbibing its values to perfect one's skills for applications of the working rule of the Upsutra.

(*v*) And to have the overview as follows

6. Here the working rule of Upsutra-1 'to follow the forms as framed', symmetrically/proportionately, shall be of real help.
7. The working rule of Ganita Sutra-1 shall be immediately unfolding 'the counting numbers'.

8. Parallel to it, would follow the skill of 'counting'.
9. The counting as it is of the rule one more than previous is naturally the counting of increasing/ascending order.
10. Parallel to it, by the rule of symmetry and proportions rule of Upsutra-1, it shall be leading to 'reverse counting' of 'decreasing/descending order'.
11. However, the one may have a pause here and take note of the situation as that this rule of 'one more than before', infact is essentially the rule of coverage 'step by step' which would mean that every step, one shall be of 'finite range coverage' attainment, but never the less every time, one can go a step ahead of it but never the less the attainment is to retain 'of finite range of steps.
12. In a way, 'infinity', as such shall be eluding at every step.
13. Therefore the whole range but for the end infinity is to be taken as attainable by the rule.
14. Parallel to ascending counting, the descending counting as well shall be having disappointment to reach 'end infinity' by the sequential 'step by step' process.
15. With it 'infinity at both ends' of counting number line are to get excluded from the counting attainments.
16. Illustratively it would be like, one may think of 'bamboo' of any number of 'steps' but it is not to take us to infinity.

## 5. Ganita Sutra-2

**SUTRA - II**

निखिलं नवतश्चरमं दशतः

| 01 | 02 | 03 | 04 | 05 | 06 | 07 | 08 | 09 | 10 |
|---|---|---|---|---|---|---|---|---|---|
| न् | इ | ख् | इ | ल् | अ | ं | न् | अ | व् |
| 11 | 12 | 13 | 14 | 15 | 16 | 17 | 18 | 19 | 20 |
| अ | त् | अ | श | च् | अ | र | अ | म् | अ |
| 21 | 22 | 23 | 24 | 25 | 26 | 27 | 28 | | |
| | द् | अ | श् | अ | त् | अ | : | | |

| Total Letters | Vowels | Nasels | Consonents |
|---|---|---|---|
| 28 | 12 | 3 | 13 |

## 6. Initial Steps Outline

(*i*) Read the text of the Sutra.

(*ii*) Pronounce the text Loudly.

(*iii*) Chase the text for its working rule

(*iv*) Think, meditate and transcend through the text of the Sutra to have complete insight and for fully imbibing its values to perfect one's skills for applications of the working rule of the Sutra.

(*v*) And to have the overview as follows

17. Ganita Sutra-2 with its working rule 'all from nine and last from ten' in a way, distinguishes the sequential coverage range of step by step approach, as of a pair of parts of which the first part being of 'steps 1 to 9' while the second part to be of 'the last counting step '10'.
18. It, as such is transiting from 'single digits/numerals range' of steps '0, 1, 2, 3, 4, 5, 6, 7, 8, 9' to double digit number '10'.
19. One may have a pause here and permit the transcending mind to transcend through the single digits steps processing of Ganita Sutra-1 to be at the 'double digit number'.
20. It is to be at a format for 'taming' an 'infinity'.
21. Further it is also a methodology of taming 'finite ranges' as well.
22. Still further, it shall be making available a mathematical technique of single digits range '0, 1, 2, 3, 4, 5, 6, 7, 8, 9' at base line to be availed for sequential steps increase at the index as $10^0, 10^1, 10^2, 10^3, 10^4, 10^5, 10^6, 10^7, 10^8, 10^9$.
23. Still further the transition from Ganita Sutra-1 to Ganita Sutra-2, as well shall be leading to transition from the line format to that of a surface format.
24. With this transition would be available a geometric format for organization of double digit numbers from '00 to 99' as along $10 \times 10$ geometric format.

25. Still further, the double digit numbers '01 to 99' would permit their re-organization as along 9 × 11 matrix format of 9 columns and 11 rows.

26. These pair of formats for re-organization of double digit numbers of ten place value system in terms of the numerals '0, 1, 2, 3, 4, 5, 6, 7, 8, 9', with the help of the symmetry/proportions rule of Upsutra-1 shall be providing the whole range of place value systems with their numerals being of single digit values and the place value to be of double digit expression.

## 7. Ganita Upsutra-2

**UPSUTRA - II शिष्यते शेषसंज्ञः**

| 01 | 02 | 03 | 04 | 05 | 06 | 07 | 08 | 09 | 10 |
|---|---|---|---|---|---|---|---|---|---|
| श् | इ | ष् | य् | अ | त् | ए | श् | ए | ष् |

| 11 | 12 | 13 | 14 | 15 | 16 | 17 | 18 |
|---|---|---|---|---|---|---|---|
| अ | स् | अ | | ज् | ञ् | अ | : |

| Total Letters | Vowels | Nasels | Consonents |
|---|---|---|---|
| 18 | 7 | 2 | 9 |

## 8. Initial Steps Outline

(*i*) Read the text of the Upsutra.

(*ii*) Pronounce the text Loudly.

(*iii*) Chase the text for its working rule

(*iv*) Think, meditate and transcend through the text of the Upsutra to have complete insight and for fully imbibing its values to perfect one's skills for applications of the working rule of the Upsutra.

(*v*) And to have the overview as follows

27. Further with the help of Ganita Upsutra-2 as that 'which remains is designated and named as remainder', shall be permitting transition from the first format of double digit numbers '00 to 99' to '01 to 99' as that 10 × 10 = 100, a three digit number 'remains outside the expression

range of '99 double digit numbers' 01 to 99 as of the format 9 × 11.

28. The above expression formats would get coordinated by the identity $10 \times 10 - 1 \times 1 = 9 \times 11$.
29. In general for n place value system this transition shall be of the identity values $n^2 - 1^2 = (n - 1)(n + 1)$.
30. One may have a pause here for having a transition from Ganita Sutra-2 to Ganita Sutra-3.
31. The transition from Ganita Sutra-1 to Ganita Sutra-2 is of the format values of transition from a line to a plane.
32. A Step ahead, the transition is to be from a plane to a pair of planes; from horizontal plane to the pair of horizontal and vertical planes.
33. As the transition from line to plane is to be from a single axis to a pair of axes of linear format/lines, so the transition ahead is to be from a single spatial dimension/plane (surface/2-space) as a dimension to a pair of spatial dimensions.
34. This, this way for the working rule of Ganita Sutra-3 shall be focusing upon the vertical plane (presuming as that the horizontal plane is already available).

## 9. Ganita Sutra-3

**SUTRA - III**

ऊर्ध्वतिर्यग्भ्याम्

| 01 | 02 | 03 | 04 | 05 | 06 | 07 | 08 | 09 | 10 |
|---|---|---|---|---|---|---|---|---|---|
| ऊ | थ् | र् | व् | अ | त् | इ | र् | य् | अ |
| 11 | 12 | 13 | 14 | 15 | | | | | |
| ग् | भ् | य् | आ | म् | | | | | |

| Total Letters | Vowels | Nasels | Consonents |
|---|---|---|---|
| 15 | 5 | - | 10 |

## 10. Initial Steps Outline

(*i*) Read the text of the Sutra.

(*ii*) Pronounce the text Loudly.

(*iii*) Chase the text for its working rule

(*iv*) Think, meditate and transcend through the text of the Sutra to have complete insight and for fully imbibing its values to perfect one's skills for applications of the working rule of the Sutra.

(*v*) And to have the overview as follows

35. The working rule of Ganita Sutra-3 'vertically and crosswise' as such while chased along the vertical plane, shall be of the features of an upward progression along the diagonal.

36. It in a way, as such, is providing a shift from side of a square/rectangle to be along the diagonal of a square/rectangle.

## 11. Ganita Sutra-4

**SUTRA - IV**

**परावर्त्य योजयेत्**

| 01 | 02 | 03 | 04 | 05 | 06 | 07 | 08 | 09 | 10 |
|---|---|---|---|---|---|---|---|---|---|
| प् | अ | र् | आ | व् | अ | र् | त् | य् | अ |
| 11 | 12 | 13 | 14 | 15 | 16 | 17 | | | |
| य् | ओ | ज् | अ | य् | ए | त् | | | |

| Total Letters | Vowels | Nasels | Consonents |
|---|---|---|---|
| 17 | 7 | - | 10 |

## 12. Initial Steps Outline

(*i*) Read the text of the Sutra.

(*ii*) Pronounce the text Loudly.

(*iii*) Chase the text for its working rule

(*iv*) Think, meditate and transcend through the text of the Sutra to have complete insight and for fully imbibing its values to perfect one's skills for applications of the working rule of the Sutra.

(*v*) And to have the overview as follows

37. One may again have a pause for attaining transition from Ganita Sutra-3 format to that of Ganita Sutra-4 format.
38. One may have a fresh look at the format of Ganita Sutra-3 with a focus upon the diagonal of square/rectangle/plane/surface which splits it into two parts.

## 13. Ganita Upsutra-3

**UPSUTRA - III**

आधमाधेनान्त्यमन्त्येन

| 01 | 02 | 03 | 04 | 05 | 06 | 07 | 08 | 09 | 10 |
|---|---|---|---|---|---|---|---|---|---|
| आ | द् | य् | अ | म् | आ | द् | य् | ए | न् |
| 11 | 12 | 13 | 14 | 15 | 16 | 17 | 18 | 19 | 20 |
| आ | न् | त् | य् | अ | म् | अ | न् | त् | य् |
| 21 | 22 | 23 | | | | | | | |
| ए | न् | अ | | | | | | | |

| Total Letters | Vowels | Nasels | Consonents |
|---|---|---|---|
| 23 | 9 | - | 14 |

## 14. Initial Steps Outline

(*i*) Read the text of the Upsutra.

(*ii*) Pronounce the text Loudly.

(*iii*) Chase the text for its working rule

(*iv*) Think, meditate and transcend through the text of the Upsutra to have complete insight and for fully imbibing its values to perfect one's skills for applications of the working rule of the Upsutra.

(*v*) And to have the overview as follows

39. Ganita Upsutra-3 with its working rule of pairing starting points amongst themselves and end points amongst themselves, shall be helping us have comparisons for the pair of diagonals of square/rectangle/plane/surface.
40. In the background of the feature of a diagonal splitting surface/plane/rectangle/square into a pair of halves as reflection pair, it may be a smooth transition from Ganita Sutra-3 to Ganita Sutra-4.

41. One shall have a pause and have a fresh look at the working rule of Ganita Sutra-4 'transpose and unite'.
42. The transposition of one part of the surface/ plane/rectangle/square and its uniting with the other part of it, is in a way is to be of reversal of the above step of Sutra-3 of splitting the vertical plane into a pair of halves.
43. It is this reversal in a way would amount reunifying of the pair of parts. Moreover it would be relevant to note that under the rule of symmetry/proportions of following the forms as framed as that either of the two diagonals of the vertical plane shall be the choice for the chase of Ganita Sutra-3 working rule and in both ways the working rule of 'Ganita Sutra-4' shall be applicable.
44. Here one may again have a pause and have a fresh look at the working rule of Ganita Upsutra-4.
45. Ganita Upsutra-4 provides as that such features are to be only uptill the order of 'artifice 7'.
46. With it, to have full comprehension there of and to have deep insight about values and virtues of 'artifice 7', it shall be appropriate to have chase of artifice 7.
47. Amongst others, the immediate features of artifice 7 which would be glaringly coming to focus would be, as that, (1) seven is the biggest prime numeral of the ten place value system (2) Of the twelve edges of the cube, only 7 are sufficient to sequentially coordinate all the eight corners of the cube. (3) cube as representative regular body of 3-space as precisely seven versions parallel to seven geometries of 3-space as of signatures range (0, 1, 2, 3, 4, 5, 6) (4) The visible spectrum is of seven colors band.
48. As such the features of Sun light are coming into play for processing from fourth step onward.
49. It is going to be a processing ahead of 3-space/triloki.
50. One may, in the circumstances shall have a pause and to think, meditate and transcend through 3-space/cube, a

linear order set up leading to sequential increasing range only up till hyper circles 1 to 7 and not beyond.

51. The decreasing sequential order for hyper cube 8 onwards is the phenomenon which is coming into play only because of the upper limit of linear order being of the range of this set ups of seven geometries of 3-space/seven version of cube.

## 15. Black Nature, Black Colour

52. The pair of axes of the horizontal plane shall be leading to pair of vertical planes. Further as each plane shall be having a pair of diagonals, and as such it shall be making quadruple structure upon the horizontal plane.
53. As above, so below of the horizontal plane as well would follow quadruple structures. And this would be parallel to octave cut for the 3-space.
54. With it the horizontal plane of middle placement for octave structures shall be permitting eight points (corner points) enveloping for the centre of the horizontal plane.
55. The centre enveloped by eight points as such shall be distinguishing itself from the eight points enveloping it.
56. One may have a pause here and have a fresh look for the transition steps starting with Sutra-1 and Upsutra-1, and sequentially being through Sutra-2 and Upsutra-2, Sutra-3 and Upsutra-3 and finally to be at Sutra-4 and Upsutra-4.
57. One way to chase these transition steps would be in terms of the number of letters availed by the texts of Sutra-1 and Upsutra-1 together as to be of the range 16 + 10 = 26 letters range.
58. Artifice 26, as such would come into play with its whole range of features and also with its split and re-organization as 26 = 16 + 10.
59. Artifice 26 amongst other features is known for its prominent roles as 26 elements range of Vishnu lok (6-space), further as 26 geometric components of geometric

envelop of cube, namely eight corner points, twelve edges and six surfaces, and still further as 26 primes (including one) up till one hundred, namely 1, 2, 3, 5, 7, 11, 13, 17, 19, 23, 29, 31, 37, 41, 43, 47, 53, 59, 61, 67, 71, 73, 79, 83, 89 and 97.

60. Ganita Sutra-1 with its first letter being the sixth vowel and second letter being the first varga consonant of the format of spatial order of 4-space and of dimension fold of the manifestation layer (2, 3, 4, 5), shall be leading to sequential flow at first step being of artifice 6 and at second step being of artifice 4, which as such provides 6 × 4 = 24, artifice which is to be of the range of artifices 1, 2, 3, 4, 5, 6, 7, 8, 9, 10, 11, 12, 13, 14, 15, 16, 17, 18, 19, 20, 21, 22, 23, 24, 25, 26' to be of the feature of the format for 24 letters alphabet (Greek alphabet) and as such these letters getting associated with the values of artifice one to artifice twenty four respectively.

61. This expression range of twenty four steps, as such together with the pair of end points for the expression range shall be extending it as to be of twenty six steps long range.

62. Ancient wisdom approach to Vishnu lok (6-space) is of 24 elements range and of 26 elements range.

63. The extension of 24 elements/alphabets letters range of Greek alphabet to 26 elements/letters range of English alphabet, as such shall be permitting association for 26 letters of English alphabet parallel to the range of artifice 1 to 26.

64. It as such shall be taken as first to twenty six letters respectively being of number value formats (in short NVFs) from 1 to 26, in that sequence and order.

65. In terms of it the word formulation (English) as such shall be accepting number value format of artifice (E = 5, N = 14, G = 7, L = 12, I = 9, S = 19 and H = 8) with summation

value 5 + 14 + 7 + 12 + 9 + 8 = 74 = NVF (PAIRING) = NVF (ELEMENT).

66. This as such shall be making ENGLISH being the language of PAIRING (ELEMENT).
67. This pairing feature is associated as well as cumulative.
68. The split of text of Ganita Sutra-1 as of a pair of formulations of artifices ranges (9, 7) when expressed (9, 7), the same simultaneously would permit expression as (7, 9).
69. NVF (NATURE) = 79 = NVF (HAPPENS) and NVF (OCCURES) = 84 = NVF (COLOUR).
70. And NVF (IT) = 29 = NVF (BLACK).
71. As such NVF (IT HAPPENS) = NVF (BLACK NATURE) and NVF (IT OCCURES) = NVF (BLACK COLOUR).
72. NVF (NATURE) = NVF (COLOUR) – NVF (E); NVF (E) = 5, and it as such shall be helping us have an insight for transition from Ganita Sutra-4 to Ganita Sutra-5.
73. Transition from Ganita Sutra-4 to Ganita Sutra-5, as such would be a transition from artifice 4 to artifice 5.
74. It would be a transition from the alphabet letter (D) to alphabet letter (E).
75. One may have a fresh look at the word formulation (SEED), and would take note as that this formulation sequentially phases itself as SeeD.
76. Ancient wisdom accepts a sequential phasing as Vyakata व्यक्ता, Avakata अव्यक्ता, Avakato avakatat अव्यक्तो अव्यक्तात, Santanta lukru and Pursha format पुरूशा फारमेट.
77. This sequential range is of 3-space, 4-space, 5-space and 6-space in that ascending sequence and order.
78. This as such shall be leading us along the format for transition from artifice 3/3-space/cube/linear order domain/Ganita Sutra-3 to four/4-space/hyper cube 4/ spatial order 4-space domain/Ganita Sutra-4.

## 16. Ganita Sutra-5

**SUTRA - V**

शून्यं साम्समुच्चये

| 01 | 02 | 03 | 04 | 05 | 06 | 07 | 08 | 09 | 10 |
|---|---|---|---|---|---|---|---|---|---|
| श् | ऊ | न् | य् | अ | ं | स् | आ | म् | य् |
| **11** | **12** | **13** | **14** | **15** | **16** | **17** | **18** | **19** | **20** |
| अ | स् | अ | म् | उ | च् | च् | अ | य् | ए |

| Total Letters | Vowels | Nasels | Consonents |
|---|---|---|---|
| 20 | 8 | 1 | 11 |

## 17. Initial Steps Outline

(*vi*) Read the text of the Sutra.

(*vii*) Pronounce the text Loudly.

(*viii*) Chase the text for its working rule.

(*ix*) Think, meditate and transcend through the text of the Sutra to have complete insight and for fully imbibing its values to perfect one's skills for applications of the working rule of the Sutra.

(*x*) And to have the overview as follows.

79. Ahead would be sequential transition formats from Ganita Sutra-4 to Ganita Sutra-5 as transition from artifice 4/hyper cube 4/4-space/spatial order 4-space domain/ Ganita Sutra-4 to artifice 5/5-space/hyper cube 5/solid order 5-space domain/Ganita Sutra-5.

80. One shall have a fresh look at the texts and working rules of Ganita Sutras 4 and 5 in that sequence and order and parallel to it as to be of Ganita Upsutras 4 and 5 respectively.

## 18. Ganita Sutra-6

**SUTRA - VI**
**( आनुरूप्ये ) शून्मन्यत्**

| 01 | 02 | 03 | 04 | 05 | 06 | 07 | 08 | 09 | 10 |
|---|---|---|---|---|---|---|---|---|---|
| आ | न् | उ | र् | ऊ | प् | य् | ए | श् | ऊ |
| 11 | 12 | 13 | 14 | 15 | 16 | 17 | 18 | 19 | |
| न् | य् | अ | म् | अ | न् | य् | अ | त् | |

| Total Letters | Vowels | Nasels | Consonents |
|---|---|---|---|
| 19 | 8 | - | 11 |

## 19. Initial Steps Outline

(*i*) Read the text of the Sutra.

(*ii*) Pronounce the text Loudly.

(*iii*) Chase the text for its working rule

(*iv*) Think, meditate and transcend through the text of the Sutra to have complete insight and for fully imbibing its values to perfect one's skills for applications of the working rule of the Sutra.

(*v*) And to have the overview as follows

81. One may have a pause here and have a fresh look at the texts of Ganita Sutras 5 and 6.
82. Both, Ganita Sutra 5 as well as Ganita Sutra-6 are about the values and features of 'Sunayam'.
83. The simple general use meaning of 'Sunayam' is zero.
84. However, the 'Cipher'/infinitsiable/negligible/smallest possible value, and 'zero' are of different values and virtues.
85. Ganita Sutra-5 goes to the base of un manifest (4-space)/ 5-space at the base of 4-space) while Ganita Sutra goes at the format of the base of unmanifest/6-space at the format of 5-space.
86. Here It would be relevant to note that Ganita Sutra-6 has part of its text a formulation (Anurupaya (आनुरूप्ये)

शून्यमन्यत्ध (*Anurupye*) *Sunyamanyat*), which literally would mean 'of similar forms' and formats.

87. One shall sit comfortably and permit the transcending mind to think, meditate and transcend through the texts of Ganita Sutras 4, 5 and 6 in that sequence and order.
88. Ganita Sutra-5 ' शून्य साम्यसमुच्चये /*Sunyam Samya-samuccaye/* If the samuccaya is the same it is Zero' shall be admitting simple English rendering for it as that Sunayam (zero), in Samaya (smooth transition) state shall be sum of equal samyachaya/rising.
89. It is as such of the values format 'zero is smooth rising of cipher values in equal bits (of dust like diminishing values particles).
90. Ganita Sutra-6 text ((आनुरूप्ये) शून्यमन्यत्/(*Anurupye*) *Sunyamanyat/*If one is in Ration the others is Zero) shall be admitting simple English rendering as that anurupeya in similarity, would Anayat follow the sunyam (0).
91. Zero is to be similarly follow the cipher.
92. Here It would be relevant to note that artifice 5/5-space being of solid dimensional order, as such the solids (at dimensional level) shall be sequentially to be of diminishing values of the order of ciphers (as of dust particles).
93. Ahead artifice 6/6-space being of hyper dimensional order/4-space in the role of dimension of 6-space), as such it shall be leading from cipher to zero.
94. It would be a lead similar to a going from 'domain to dimension' to reach at 'from dimension to dimension of dimension.
95. In the context of 6-space as domain to 4-space as its dimensional order, reach at a step ahead, similarly would be a going from '4-space as dimension to 2-space as dimension of dimension' which at the second sequential step would be leading from 2-space to 0-space as its dimension.

96. Here one may have a pause and recapitulate the different features, values and virtues of '0', as a previous state which under the rule of Sutra-1 to take to '1'

97. Further at the phase and stage of Ganita Upsutra-2, '0' shall be playing the role of a place value 'there, as ten place value'; but in general as of any place value.

98. Still further it is to play its role at the index, parallel to its role at the base as $n^0 = 1$, and here n can be a randam but finite place value and as such it shall be taking care of even of n square, n cube, and like that $n^m$ place values for all finite and n and m.

99. Still further it is to take from Vyakata (manifest) to unmanifest Avakata and also to Avakato Avakatat as well as to Pursha format values.

100. This, this way shall be bringing on face to face with the blissful unfolding of Sunyam for its inner folds which in a way shall be making it a '0-space',

101. It is, as such a blissful range of phases from cipher to zero and ahead from zero to 0-space.

102. 0-space like all other finite order spaces shall be playing its roles of different folds formats of manifestation layers, of transcendental ranges, self referral states and so on.

103. 0-space as domain shall be of the features of 0-space content.

104. Zero space with its placement between (–1) space and (+ 1) space shall be making it a space of continuity values and features.

105. This role of 0-space as being of continuity of features is there because of such features of zero space domain and of zero space content.

106. The following 4 × 4 matrix format for different roles of dimensional spaces, in the context of the roles of 0-space, shall be of the features as follows.

| | | | |
|---|---|---|---|
| –3 | –2 | –1 | 0 |
| –2 | –1 | 0 | 1 |
| –1 | 0 | 1 | 2 |
| 0 | 1 | 2 | 3 |

107. It would be blissful exercise to permit the transcending mind to be face to face with different roles of 0-space and to glimpse and imbibe its values as 0-space being origin, 0-space being domain, 0-space being boundary and 0-space being dimensional order.

108. One shall to fulfil the intensity of urge about the values and virtues of 'zero', shall remain in prolonged sittings of trans, as many number of times, as it is going to be blissful for the sadkhas.

## 20. Processing at Dimension of Dimension Level

109. 4-space, because of its spatial dimensional order, shall be taking to 0-space at dimension of dimension level. With 0-space in the role of dimension of 2-space and being in the role of dimension of dimension of 4-space shall be symmetrically leading to (–1) space as dimension 0f ( + 1) space and further being the dimensional of dimension of 3-space,

## 21. Ganita Sutra-7

**SUTRA - VII**

संकलव्यवकलनाभ्याम्

| 01 | 02 | 03 | 04 | 05 | 06 | 07 | 08 | 09 | 10 |
|---|---|---|---|---|---|---|---|---|---|
| स् | अ | | क् | अ | ल् | अ | न् | अ | व् |
| 11 | 12 | 13 | 14 | 15 | 16 | 17 | 18 | 19 | 20 |
| य् | अ | व् | अ | क् | अ | ल् | अ | ल् | आ |
| 21 | 22 | 23 | 24 | | | | | | |
| भ् | य् | आ | म् | | | | | | |

| Total Letters | Vowels | Nasels | Consonents |
|---|---|---|---|
| 24 | 10 | 1 | 13 |

## 22. Initial Steps Outline

(*i*) Read the text of the Sutra.

(*ii*) Pronounce the text Loudly.

(*iii*) Chase the text for its working rule

(*iv*) Think, meditate and transcend through the text of the Sutra to have complete insight and for fully imbibing its values to perfect one's skills for applications of the working rule of the Sutra.

(*v*) And to have the overview as follows

110. It is this simultaneous availability of the processing formats of + 1-space and of (–1) space shall be processing format/working rule of Ganita Sutra-7 'संकलनव्यवकलनाभ्याम् / *Sankalana-vyavakalanbhyam*/By addition and by subtraction'.

111. Simultaneous addition (along + 1-space and –1-space), as such would amount to simultaneous availing of the positive orientation, as well as of negative orientation of a line.

112. One may have a pause here and have a fresh look at the set up of a line of pair of orientations.

113. The line as a format of a moving point is vesting the line format with a pair of orientations because of the fact that 0-space plays the role of dimension of 2-space/plane.

114. This, this way coordinates the range of consecutive dimensional spaces (–1) space, 0-space, + 1-space and 2-space.

115. It is this feature of a line/interval as the representative regular body of 1-space shall be of the manifestation format of spatial order 4-space.

116. Sequentially the set up of interval/line as of a manifestation format within spatial order creator space (4-space) shall be of the features as that here (–1) space plays the role of dimension of + 1-space, while 0-space plays the role of boundary of 1-space, as well as of the role of dimension of 2-space.

117. However, 2-space, as such shall be playing the role of origin of line/interval being manifestation layer.
118. This, this way shall be bringing to focus the feature of the transition format for the dimensional bodies (in particular that of 1-space body) being of a four fold manifestation layers, which in the case of 1-space body/line/interval shall be of four fold expression (–1) space, 0-space, + 1-space, 2-space, and in general n space body, to be designated as hyper cube n, shall be a manifestation layer of four fold expression (n–2) space, (n–1) space, n space, and (n–1) space for all integrals values of n.

23. **Ganita Upsutra-7**

**UPSUTRA - VII**

**यावदूनं तावदूनीकृत्य वर्गं च योजयेत्**

| 01 | 02 | 03 | 04 | 05 | 06 | 07 | 08 | 09 | 10 |
|---|---|---|---|---|---|---|---|---|---|
| य् | आ | व् | अ | द् | ऊ | न् | अ | | त् |
| 11 | 12 | 13 | 14 | 15 | 16 | 17 | 18 | 19 | 20 |
| आ | व् | अ | द् | ऊ | न् | ई | क् | ऋ | त् |
| 21 | 22 | 23 | 24 | 25 | 26 | 27 | 28 | 29 | 30 |
| य् | अ | व् | अ | र् | ग् | अ | | च् | अ |
| 31 | 32 | 33 | 34 | 35 | 36 | 37 | | | |
| य् | ओ | ज् | अ | य् | ए | त् | | | |

| Total Letters | Vowels | Nasels | Consonents |
|---|---|---|---|
| 37 | 16 | 2 | 19 |

**119. One may have a pause here and have a fresh look at Upsutra 7 'यावदूनम् तावदूनीकृत्य वर्ग च योजयेत्/ *Yavadunam Tavadunikrtya Varganca Yogayet/*That twice This twice Square and add'.**

**120. Here It would be relevant to note that the double digit artifices pair (10, 01) constitute a reflection pair; it, as such brings into reflection processing features responsible for pair of orientations for image and object with mirror at their middle.**

121. It is the 0 value association with the middle placement for the mirror, as such shall be splitting the object – image range as to be of a pair of halves of opposite orientations, and this as such shall be the emerging format instead of the increasing line format or decreasing line format.

## 24. Ganita Sutra-8

**SUTRA - VIII**

**पूरणापूरणाभ्याम्**

| 01 | 02 | 03 | 04 | 05 | 06 | 07 | 08 | 09 | 10 |
|---|---|---|---|---|---|---|---|---|---|
| प् | ऊ | र् | अ | ण् | आ | प् | ऊ | र् | अ |
| 11 | 12 | 13 | 14 | 15 | 16 | | | | |
| ण् | आ | भ् | य् | आ | म् | | | | |

| Total Letters | Vowels | Nasels | Consonents |
|---|---|---|---|
| 16 | 7 | - | 9 |

## 25. Initial Steps Outline

(*i*) Read the text of the Sutra.

(*ii*) Pronounce the text Loudly.

(*iii*) Chase the text for its working rule

(*iv*) Think, meditate and transcend through the text of the Sutra to have complete insight and for fully imbibing its values to perfect one's skills for applications of the working rule of the Sutra.

(*v*) And to have the overview as follows

122. It is this synthetic feature of the emerging format which shall be governing the further transition features from Ganita Sutra-7 to Ganita Sutra-8 onwards.

123. One may again have a pause and to have a fresh look at Ganita Sutra-8 'पूरणपूरणाभ्याम्/ *Puranapuranabhyan*/By the completion or non-completion'

124. This transition from synthetic format of Ganita Sutra-7 to that of Ganita Sutra-8 as such, firstly while shall be generalizing the synthetic format it shall be making the middle/joint as to be of fluctuating features.

125. The fluctuating middle/joint of the closed interval (or of open interval) or of half open interval shall be bringing

to focus the features of a point as to be of 0-space, and of to be '1-space', as being of different structural set ups.

126. It is in the context, it would be possible to comprehend as to how the half closed interval/half open interval distinguishing itself from that of closed interval merely in the light of 'absence or presence' of a single point.

127. Such a prominent role of a 'point' which structurally distinguishes the set up of half closed interval from closed interval and of half open interval from open interval.

128. One may again have a pause and to have a fresh look at set ups of closed interval and of open interval.

129. These shall be distinguishing themselves from each other in terms of a pair of points.

130. It, this way, takes us from the presence or absence of a single point to that of the presence or absence of a pair of points.

131. This feature, as such shall be focusing upon the role of a fluctuating middle point.

132. This also shall be focusing upon the presence or absence of both are of a single n point of an interval.

26. **Ganita Sutra-9**

**SUTRA - IX**

चलनकलनाभ्याम्

| 01 | 02 | 03 | 04 | 05 | 06 | 07 | 08 | 09 | 10 |
|---|---|---|---|---|---|---|---|---|---|
| च् | अ | ल् | अ | न् | अ | क् | अ | ल् | अ |
| 11 | 12 | 13 | 14 | 15 | 16 | | | | |
| न् | आ | भ् | य् | भ् | म् | | | | |

| Total Letters | Vowels | Nasels | Consonents |
|---|---|---|---|
| 16 | 7 | - | 9 |

## 27. Initial Steps Outline

(*i*) Read the text of the Sutra.

(*ii*) Pronounce the text Loudly.

(*iii*) Chase the text for its working rule

(*iv*) Think, meditate and transcend through the text of the Sutra to have complete insight and for fully imbibing its values to perfect one's skills for applications of the working rule of the Sutra.

(*v*) And to have the overview as follows

133. This, this way shall be leading to transition from the format of Ganita Sutra-8 to that of Ganita Sutra-9.
134. Ganita Sutra-9 'चलनकलनाभ्याम्/ *Calana-kalanabhyam/* Differen-tiation Calculus', brings to focus the computation processes for higher domains wrapped within lower domains in general and 1-space body/interval accepting 0-space/points as n point.
135. Further, specifically it focuses upon the fluctuating feature of the in between point.
136. This as such because of the in between point being of middle point features, as such it operates along the specific orientations formats.
137. Because of it, it aims to approach of the boundary a single n point.
138. It is this feature of the fluctuating middle point which at a time ensures only the coverage of half of the boundary (a single n point).
139. In the process the second half boundary/the second n point is bound to remain uncovered.
140. It is this feature of such processing in terms of fluctuating middle point which deserves to be fully comprehended and to be completely imbibed as other wise the continuity and derivability are to go distinctive ways.
141. To sustain continuity and derivability features to be parallel throughout the domain would as such require that both orientations formats for the fluctuating middle point are to be taken account of.

28. Ganita Sutra-10

**SUTRA - X**

यावदूनम्

| 01 | 02 | 03 | 04 | 05 | 06 | 07 | 08 | 09 |
|---|---|---|---|---|---|---|---|---|
| य् | आ | व् | अ | द् | ऊ | न् | अ | म् |

| Total Letters | Vowels | Nasels | Consonents |
|---|---|---|---|
| 9 | 4 | - | 5 |

## 29. Initial Steps Outline

(*i*) Read the text of the Sutra.

(*ii*) Pronounce the text Loudly.

(*iii*) Chase the text for its working rule

(*iv*) Think, meditate and transcend through the text of the Sutra to have complete insight and for fully imbibing its values to perfect one's skills for applications of the working rule of the Sutra.

(*v*) And to have the overview as follows

142. For this, the transition is to be had from Ganita Sutra-9 to Ganita Sutra-10.

143. Ganita Sutra-10 'यावदूनम् *Yavadunam*/By the Deficiency' aims to take care of both orientations of a fluctuating point.

144. One may again have a pause and permit the transcending mind to comprehend and imbibe the full range of features of the fluctuating middle point.

145. It shall be bringing to focus as that the part shall be making itself congruent to the whole because of any interval, how so ever small it be is to accommodate infinite points (of 0-space value).

146. Illustratively every pair of un equal lines shall be having infinite set of points, and every pair of surfaces, as well shall be accommodating infinite points and ahead every pair of solids shall be accommodating infinite points.

## 30. Ganita Sutra-11

**SUTRA - XI**

**व्यष्टिसमिष्टिः**

| 01 | 02 | 03 | 04 | 05 | 06 | 07 | 08 | 09 | 10 |
|---|---|---|---|---|---|---|---|---|---|
| व् | य् | अ | ष् | ट् | इ | स् | अ | म् | अ |
| 11 | 12 | 13 | 14 | | | | | | |
| प् | ट् | इ | : | | | | | | |

| Total Letters | Vowels | Nasels | Consònents |
|---|---|---|---|
| 14 | 5 | 1 | 8 |

## 31. Initial Steps Outline

(*vi*) Read the text of the Sutra.

(*vii*) Pronounce the text Loudly.

(*viii*) Chase the text for its working rule

(*ix*) Think, meditate and transcend through the text of the Sutra to have complete insight and for fully imbibing its values to perfect one's skills for applications of the working rule of the Sutra.

(*x*) And to have the overview as follows

147. This feature, as such shall be helping us to comprehend and imbibe the features of Ganita Sutra-11 'व्यश्टिसमश्टि/ *Vyastisa-mastih*/Specific and General'.
148. Let us again have a fresh look at the fluctuating nature of the middle point.
149. It would bring to pointed attention as that the middle point, as such, acquires new structural features.
150. It is a phenomenon which in ancient wisdom is designated and is known as Divya Ganga flow from within at the level of the base of the manifestations.
151. It is parallel to the transcendence phenomenon from the Bindu Sarovar (of sole syllable Om (ॐ) ).
152. The parallel phenomenon is the ascendance phenomenon.
153. It is from the transcendental base (5-space) of the origin fold of the manifestations (4-space).

154. Srimad Bhagwad Geeta at the end of its each of the eighteen chapters (of Yoga Nishtha aspects) accepts pushpika/colophon which is fulfilled and is blossoming with the transcendental values flowing as a Divya Ganga flow into the transcendental base.

155. The organization format of Samved Samhita as well avails this ascendance process of transcendental values from within the transcendental base (5-space) into creator space (4-space).

156. This, as such will help the sadkhas comprehend and chase the Ganita Sutras format from Ganita Sutra-11 to Ganita Sutra-1 as 11 steps chase parallel to 11 versions of hyper cube 5 as dimensional bodies of 11 distinct geometries of 5-space.

## 32. Ganita Sutra-12

**SUTRA - XII**

शेषाण्यङ्केन चरमेण

| 01 | 02 | 03 | 04 | 05 | 06 | 07 | 08 | 09 | 10 |
|---|---|---|---|---|---|---|---|---|---|
| श् | ए | ष् | आ | ण् | य् | अ | ङ् | क् | ए |
| 11 | 12 | 13 | 14 | 15 | 16 | 17 | 18 | 19 | 20 |
| न् | अ | च् | अ | र् | अ | म् | ए | ण् | अ |

| Total Letters | Vowels | Nasels | Consonents |
|---|---|---|---|
| 20 | 9 | - | 11 |

## 33. Initial Steps Outline

(*i*) Read the text of the Sutra.

(*ii*) Pronounce the text Loudly.

(*iii*) Chase the text for its working rule

(*iv*) Think, meditate and transcend through the text of the Sutra to have complete insight and for fully imbibing its values to perfect one's skills for applications of the working rule of the Sutra.

(*v*) And to have the overview as follows

157. It ascendance ahead from Ganita Sutra-11 to Ganita Sutra-12, as such will be in terms of the 'residue/ reminder' of uncovered domain ahead.
158. It as such is going to be the 12th phase and stage ahead of 11 versions of hyper cube 5.
159. The 12th phase and stage for the hyper cube 5/5-space is of the organization features of 12 components of the transcendental boundary of self referral domain (6-space)/hyper cube 6.
160. One may have a pause here and permit the transcending mind to think, meditate and to transcend through the text of Ganita Sutra-12 ' शेषण्यड्.केन चरमेण/*Sesnyankena Caramena*/The Remainder by the last digit'.
161. The 12th component/state of transcendental (5-space) domain/hyper cube 5 is the last/final component of the transcendental boundary.
162. One may again have a pause and permit the transcending mind to remain in prolonged deep sittings of trans and to glimpse the existence phenomenon of a phase and stage of self referral domain within transcendental boundary.
163. It is the phase and stage of of 'viloknam' being face to face with the self referral existence (of 6-space/Pursha format).

## 34. Ganita Upsutra-12

**UPSUTRA - XII**

**विलोकनम्**

| 01 | 02 | 03 | 04 | 05 | 06 | 07 | 08 | 09 |
|---|---|---|---|---|---|---|---|---|
| व् | इ | ल् | ओ | क् | अ | न् | अ | म् |

| Total Letters | Vowels | Nasels | Consonents |
|---|---|---|---|
| 9 | 4 | - | 5 |

**164. It would be the phase and stage of Ganita Upsutra-12 'विलोकनम/ *Vilokanam*/By observation'.**

35. Ganita Upsutra-13

**UPSUTRA - XIII**

**गुणितसमुच्चय: समुच्चयगुणित**

| 01 | 02 | 03 | 04 | 05 | 06 | 07 | 08 | 09 | 10 |
|---|---|---|---|---|---|---|---|---|---|
| ग् | उ | ण् | इ | त् | अ | स् | अ | म् | उ |
| 11 | 12 | 13 | 14 | 15 | 16 | 17 | 18 | 19 | 20 |
| च् | च् | अ | य् | अ | : | स् | अ | म् | उ |
| 21 | 22 | 23 | 24 | 25 | 26 | 27 | 28 | 29 | 30 |
| च् | च् | अ | य् | अ | ग् | उ | ण् | इ | त् |
| 31 | 32 | | | | | | | | |
| अ | : | | | | | | | | |

| Total Letters | Vowels | Nasels | Consonents |
|---|---|---|---|
| 32 | 14 | 2 | 16 |

165. One shall again have a pause and to permit the transcending mind to remain in prolonged deep sittings of trans to glimpse and face to face with the next phase and stage of the values and features of the Ganita Sutra-13 and Ganita Upsutra-13.

166. Ganita Sutra-13 is of the values and features ; 'सोपान्त्यद्वयमन्त्यम्/ *Sopantyadvyamantyam* /The ultimate and twice the penulti-mate'.

167. Ganita Upsutra-13 is of the values and features 'गुणितसमुच्चयः समुच्चयगुणितः / *Gunita Samuccaya Samuccaya-gunitah* /Product samuchya Samuchya Product'.

168. For proper comprehension and imbibing of the values and features of this phase and stage of Ganita Sutra-13 and ganita Upsutra-13 one may have a fresh look at the values and features of Ganita Sutra-7 and of Ganita Upsutra-7, and to sequentially re-chase the transition process as it is taking us uptill the phase and stage of Ganita Sutra-13 and Ganita Upsutra-13.

169. It shall be bringing to focus that the simultaneous 'addition and minus' features shall be peeling of the pair of orientations of a line/interval.
170. It also shall be simultaneously focusing static state of a point along with a pair of dynamic states of a moving point.
171. It is this static position together with pair of dynamic positions of a point (0-space)/ 0-space body/0-space content lump as a triple/trio/three fold expression (–1) space, 0-space, (+ 1) space] (–1, 0, + 1)/(negative orientation, dynamic state, static zero state, and positive orientation dynamic state.
172. This (trio –1, 0, + 1) in its generality (n–1), n, n + 1), shall be making it a three fold (MONAD).
173. It is because of its acquisition of fourth (origin fold) as along a manifestation format of spatial order (2-space in the role of dimension) of creator space (4-space), it shall be as a quadruple (–1, 0, 1, 2 with 2-space in the role of origin, shall be permitting a reversal of orientations at the middle of the monad (–1, 0, + 1)/ (n–1, n, n + 1) making it to be (1, 0, –1)/(n + 1, n, n–1).
174. It in a way shall be permitting a bend at the middle as both end folds, namely of values (n–1) and (n + 1) shall be super imposing themselves, either side of the middle.
175. With it the end points would emerge as of values [(n–1) + (n + 1)] = 2n.
176. It is this feature, which imbeds end points with double (2n) the value of the middle (n), which shall help comprehend and imbibe the values and features of Ganita Sutra–13 'सोपान्त्यद्वयमन्त्यम्/*Sopantyadvyamantyam*/The ultimate and twice the penultimate'.
177. One may again have a pause here and permit the transcending mind to remain in prolonged deep sittings of trans to simultaneously comprehend and imbibe the

values and virtues of Ganita Sutra-13 as well as of Ganita Upsutra-13.

178. One way to approach would be to approach Ganita Upsutra-13 in the light of the values and features of Ganita Sutra-13.

179. The other way to approach would be to comprehend and imbibe the values and virtues of Ganita Sutra-13 in the light of the values and features of Ganita Upsutra-13.

180. One value and feature of Ganita Sutra-13 is that it admits parallel but of opposite orientation of equal steps from the middle (n) to either end of value (2n).

181. One valuable feature of Ganita Upsutra-13 is that it is a composition of pair of formulations (समुच्चयगुणितः/ *Samuccayagunitah*) and (गुणितसमुच्चयः/ *GunitaSamuccaya*), which shall be taken as a reflection pair formulations.

182. The formulation (समुच्चयगुणितः/ *Samuccayagunitah*) on it sequential chase from (समुच्चय/ *Samuccay*) to (गुणितः/ *gunitah*) shall be leading to as that, the 'equal rising features' (समुच्चय/ *Samuccay*), are to be the features which are to be peeled off/segrigated/removed (gunita).

183. The complementary step of (गुणितसमुच्चयः/ *Gunita-Samuccaya*), shall be, as that the features which stand peeled off/segregated/removed (gunita) are to be revalued for their equality rising applications (समुच्चय/ *Samuccay*).

184. This, this way, in the context of Ganita Sutra-13 would be as going from n to 2n and also from 2n to n, as a similarity/proportionate symmetry of following the forms as those are framed and thereby having extension and transition from Ganita Sutra-12 read with Ganita Upsutra-12.

185. Here again one may have a pause and to permit the transcending mind to remain in deep sittings of trans to

glimpse the values and features of organization format of sixteen sutras stream range and parallel to it thirteen Ganita Sutras stream.

186. First of all, here the values and features of artifices 16 and 13 are to be taken in account.

187. Artifices 16 and 13 permit simultaneous re-organizations as 5 + 6 + 5 and 4 + 5 + 4.

188. Parallel to it would be the expression ranges of 6-space (domain) within 5-space as end points, for artifice 16 = 5 + 6 + 5 and as of 5-space domain within 4-space end points for the artifice 13 = 4 + 5 + 4.

189. These together shall be leading us to self referral domain (6-space)/hyper cube 6) bounded within transcendental boundary (5-space)/hyper cube 4.

190. This, this way as such shall be taking us to the start with position of processing of this domain (6-space) beginning with the 6th vowel (,) as the first letter of the text of Ganita Sutra-1 itself.

191. Further It would be relevant to note that the expressions ranges 5 + 6 + 5 and 4 + 5 + 4 are of three point fixations (at middle) and at both ends).

192. Therefore the difference between the expression ranges of artifices 16 and 13, being 16–13 = 3, as such, are to be comprehended, imbibed and to be approached in that light.

193. As in the process at the middle of the range of Ganita Sutras 1 to 13, i.e. at the phase and stage of Ganita Sutra-7, as being at the middle of the range, there is successful peeling off of pair of orientations of a line, as such the negative orientation which had come in independent existence, would be available for its pure and applied values.

## 36. Ganita Sutra-14

**SUTRA - XIV एकन्यूनेन पूर्वेण**

| 01 | 02 | 03 | 04 | 05 | 06 | 07 | 08 | 09 | 10 |
|---|---|---|---|---|---|---|---|---|---|
| ए | क् | अ | न् | य् | ऊ | न् | ए | न् | अ |
| 11 | 12 | 13 | 14 | 15 | 16 | 17 | | | |
| प् | ऊ | र् | व् | ए | ण् | अ | | | |

| Total Letters | Vowels | Nasels | Consonents |
|---|---|---|---|
| 17 | 8 | - | 9 |

## 37. Initial Steps Outline

(*vi*) Read the text of the Sutra.

(*vii*) Pronounce the text Loudly.

(*viii*) Chase the text for its working rule

(*ix*) Think, meditate and transcend through the text of the Sutra to have complete insight and for fully imbibing its values to perfect one's skills for applications of the working rule of the Sutra.

(*x*) And to have the overview as follows

194. It is this attainment of the Ganita Sutras 1 to 13 and Ganita Upsutras 1 to 13 ranges, which as such becomes the organization feature ahead at the base and format of the text of Ganita Sutra-14 'एकन्यूनेन पूर्वेण/ *Ekanyunena Purvena/* By One less than the One Before'.

195. As both orientations, positive orientations and negative orientations, run parallel to each other and are attainable by increasing sequential steps and decreasing sequential steps, as such both would permit parallel attainments in terms of the rule of symmetry proportions of following the form as those are framed under Ganita Upsutra-1, as such it may be taken as that the Upsutras range from Ganita Upsutra 1 to 13 comes into its repeated play from Ganita Sutra-13 upward as Ganita Upsutras 1, 2 and 3, and so on.

196. As such Ganita Sutra-14 and Ganita Upsutra-1 would deserve to be comprehended and imbibed simultaneously. It also would be relevant to note that because of the features of Ganita Sutra-1 and Ganita Sutra-2 being of opposite orientations, and as such like Ganita Upsutra-1 coming for its replay from Ganita Sutra-14, the Ganita Sutra-1 and Ganita Sutra-14 as well would be simultaneously coming into play.

197. this in a way would bring the simultaneous features of Ganita Sutra-1 and of Ganita Upsutra-1 to be simultaneously available for parallel play along with the values and features of Ganita Sutra-14.

## 38. Ganita Sutra-15

**SUTRA - XV गुणितसमुच्चयः**

| 01 | 02 | 03 | 04 | 05 | 06 | 07 | 08 | 09 | 10 |
|---|---|---|---|---|---|---|---|---|---|
| ग् | उ | ण् | इ | त् | अ | स् | अ | म् | उ |
| 11 | 12 | 13 | 14 | 15 | 16 | | | | |
| च् | च् | अ | य् | अ | : | | | | |

| Total Letters | Vowels | Nasels | Consonents |
|---|---|---|---|
| 16 | 7 | 1 | 8 |

## 39. Initial Steps Outline

(*i*) Read the text of the Sutra.
(*ii*) Pronounce the text Loudly.
(*iii*) Chase the text for its working rule
(*iv*) Think, meditate and transcend through the text of the Sutra to have complete insight and for fully imbibing its values to perfect one's skills for applications of the working rule of the Sutra.
(*v*) And to have the overview as follows

198. A step ahead naturally, it can be logically inferred as that the simultaneous availability of the features of Ganita Sutra-2 and Ganita Upsutra-2 would be available for simultaneous play along with the values and features of Ganita Sutra-15.

40. Ganita Sutra-16

**SUTRA - XVI गुणकसमुच्चय:**

| 01 | 02 | 03 | 04 | 05 | 06 | 07 | 08 | 09 | 10 |
|---|---|---|---|---|---|---|---|---|---|
| ग् | उ | ण् | अ | क् | अ | स् | अ | म् | उ |
| 11 | 12 | 13 | 14 | 15 | 16 | | | | |
| च् | च् | अ | य् | अ | : | | | | |

| Total Letters | Vowels | Nasels | Consonents |
|---|---|---|---|
| 16 | 7 | 1 | 8 |

## 41. Initial Steps Outline

(*i*) Read the text of the Sutra.

(*ii*) Pronounce the text Loudly.

(*iii*) Chase the text for its working rule

(*iv*) Think, meditate and transcend through the text of the Sutra to have complete insight and for fully imbibing its values to perfect one's skills for applications of the working rule of the Sutra.

(*v*) And to have the overview as follows

199. Likewise the simultaneous features of Ganita Sutra-3 and Ganita Upsutra-3 shall be available for simultaneous play with the values and features of Ganita Sutra-16.

200. However, here, one may again have a pause and permit the transcending mind to remain in prolonged sittings of trans to comprehend and imbibe the values and features of Ganita Sutra-15 and 16 in the sequential order.

201. Ganita Sutra-15 'गुणितसमुच्चय: / *Gunitasamuccayah*/The product of the Sum' in simple language sequentially is of the steps as that the features which are peeled off (gunita), are to be reset of equal/parallel rising (samuchaya).

202. Ahead, Ganita Sutra-16 'गुणकसमुच्चय: / *Gunaksamuccayah*/ All the Multipliers' on its sequential chase may be expressed in a simple language as that the feature which are embedded in first varga consonant 'क्'/spatial order creator space (4-space) (गुणक) are to be reset for their

equal/parallel rising within transcendental (5-space) domain (समुच्चय:).

203. This, as such takes us to phase and stage of the transcendental origin (5-space) as origin of creator space (4-space).

204. This is the blissful phase and stage of transcendental values flowing into the creator space (4-space).

205. It is the blissful state of structured point of 4-space getting fulfilled with the transcendental values (of 5-space) getting super imposed upon manifestation values of 4-space.

206. It is this superimposition of the transcendental values of solid order of 5-space upon the manifestation values of spatial order of creator space (4-space) which brings to focus the phenomenon of fluctuating middle point as a structured point monad as of three fold set up (–1, 0, 1)/(n–1, n, n + 1).

207. It is because of this that NVF (MONAD) = 47 = NVF (MIDDLE), NVF (DI MONAD) = 60 = NVF (FOUR) and NVF (TRI MONAD) = NVF (MONAD, MONAD) = NVF (MIDDLE, MIDDLE).

208. Further it would be relevant to note that it is (–3) space as a dimensional order, as that (–1) space domain is attainable.

209. Still further as that it is because of linear order (1-space in the role of dimension) that 3-space domain is structured.

210. That way a jump from (–3) space to (–1) space and jump from 1-space to 3-space is bridged by the jump from (–1) space to ( + 1) space.

211. It is the attainment of the point as a structured point of three folds/monad which shall be the attainment in the process.

212. 47 + 3 = 50 and 47–3 = 44, and like that the whole range becomes chase able with structured points.

213. Artifice 3/3-space/cube/linear order domain with inherent features of 3 = 1–(–1) + 1, and (–1) space in the role of dimension of dimension of 3-space shall be helping us have an insight as that the stitching of pair of linear dimensions is to be had in terms of the dimension of dimension of 3-space i.e. in terms of (–1) space/artifice (–1).

214. In general, (n + 2) space with n space in the role of dimension shall be requiring a dimensional stitch of (n–2) space because of which the pair of dimensions of structural values of (n) space shall be yielding the structural values as n + n–(n–2) = n + 2.

215. Here it would be relevant to note that our present day approach to dimensional spaces, as such is unconsciously is of n dimensions frame for n dimensional space but infact only pair of dimensions, as such, and that to of (n–2) structural order are marking their presence and role.

216. The moment, the third dimension comes into play it shall be always be leading to 6-space.

217. And it is this wonder which is the blissful basis of Vedic mathematics of Ganita Sutras.

218. To feed the curiosity, the stitching of three dimensions is being steps wise chased as follows:

(*i*) Step 1. The space is N space.

(*ii*) Step 2. Dimensional order of N space is of (N–2) space.

(*iii*) Step 3. Pair of dimensions would require a single dimensional stitch of dimension of dimension order i.e. of (n–4) space.

(*iv*) Stitching mathematics of pair of dimensions as such would be :

$$(N-2) + (N + 2) - (N-4) = N$$

(*v*) Stitching of third dimensions would require a pair of dimensional stitches of (N–4) order.

(*vi*) The Stitching mathematics of third dimension with a pair of stitched dimensions shall be of following steps.

(*i*) Value of stitched pair of dimensions = N

(*ii*) Contribution value of third dimensions

$$= N - 2$$

(*iii*) Value needed for pair of dimensional stitches required for third dimension to be stitched with the previous pair of dimensions

$$= 2\,(N - 4)$$

(*iv*) As such, the final value would be

$$N + (N-2) - 2\,(N-4)$$

$$= 2N - 2 - 2N + 8$$

$$= 6$$

219. It would be blissfully taking us back to the starting position of artifice 6/6th vowel/6-space/hyper cube 6 as the beginning point of the format of the very first letter of the text of very first sutra.

❑ ❑ ❑

# 70

# OVERVIEW OF GANITA UPSUTRAS DOMAIN ONE WAY TO REACH AT THEIR VALUES AND VIRTUES

1. Introductory

1. Credit goes to Jagadguru Swami Sri Bharti Krsna Tirthaji Maharaj to focus the attention of the modern world of mathematics about the potentialities of the Ganita Sutras (and Upsutras).
2. Further Swamiji has done a great service to the cause of Ganita Sutras by sharing with the sadkhas as to the way his holiness could lay hand at the keys to the meanings of these Sutras.
3. Many scholars are carrying the Discipline forward by taking clues from the introductory volumme under the title 'Vedic mathematics or Sixteen Simple Mathematical Formulae from the Vedas' edited by Sh. V. S. Aggarwal, M.A., PH.D., D. LITT., General editor, Hindu Vishva-vidyalaya, Nepal Rajya Sanskri Graanthmala Series' published by Banaras Hindu University, Varanasi on March 17, 1965'.
4. However, the fact remains that though the applications of the Sutras to the extent of the clues of the above book are being popularized but the Discipline, as such is not

unfolding ahead as the 'way his holiness Swami Sri Bharti Krsna Tirthaji Maharaj availed for reaching at the keys to the meanings of Sutras', as such, is being is not pilgrimaged with the result that 'further clues' are eluding.

5. It is on record that Swami Ji gave a talk and demonstration on Feb. 19, 1958 to a small group of student mathematicians at the California Institute of technology, Pasadena, California, during which he shared he had gone to the forest sought solitude of the forest meditation and tried to make some meanings out of the Ganita Sutras. And after long years and years of meditation in the forest, he took the help of lexicographies, lexicons of earlier times, and found to his astonishment and gratification that the Sutras dealt with mathematics, in all its branches.

6. This, as such, is the path of meditation followed by Swami Ji, and the sadkhas following this path may urge for the applications domain of the Ganita Sutras (and Upsutras). The applications of the Sutras to the then prevalent branches of known mathematics of the times of Swami Ji, have been well indicated in the literature of Swami ji reaching us. In the words of Swami ji:

   (*i*) The Sutras (aphorisms) apply to and cover each and every part of each and every chapter of each and every branch of mathematics (including arithmetic algebra, geometry—plane and solid, trigonometry—plane and spherical, conics—geometrical and analytical, astronomy, calculus — differential and integral etc., etc. In fact, there is no part of mathematics, pure of applied, which is beyond their jurisdiction;

   (*ii*) The Sutras are easy to understand, easy to apply and easy to remember; and the whole work can be truthfully summarized in one word "mental"!

   (*iii*) Even as regards complex problems involving a good number of mathematical operations (consecutively or even simultaneously to be performed), the time taken by the Vedic method will be a third, a fourth, a tenth or even a much smaller fraction of the time required according to modern (i.e. current) Western methods).

(*iv*) And, in some very important and striking cases, sums requiring 30, 50, 100 or even more numerous and cumbrous "steps" of working (according to the current Western methods) can be answered in a single and simple step of work by the Vedic method! And little children (of only 10 or 12 years of age) merely look at the sums written on the blackboard (on the platform) and immediately shout out and dictate the answers from the body of the convocation hall (or other venue of the demonstration). And this is because, as a matter of fact, each digit automatically yields its predecessor and its successor! And the children have merely to go on tossing off (or reeling off) the digits one after another (forwards or backwards) by mere mental arithmetic (without needing pen or pencil, paper or slate etc)!

(*v*) On seeing this kind of work actually being performed by the little children, the doctors, professors and other "big-guns" of mathematics are wonder struck and exclaim: — "Is this mathematics or magic"? And we invariably answer and say: "It is both. It is magic until you understand it; and it is mathematics thereafter"; and then we proceed to substantiate and prove the correctness of this reply of ours! And

(*vi*) As regards the time required by the students for mastering the whole course of Vedic mathematics as applied to all its branches, we need merely state from our actual experience that 8 months (or 12 months) at an average rate of 2 or 3 hours per day should suffice for completing the whole course of mathematical studies on these Vedic lines instead of 15 or 20 years required according to the existing systems of the Indian and also of foreign universities."

7. However, these sutras being of Vedic order, as such there applications cannot taken to be of applications within the prevalent mathematics domains only. The real functional domain of the values and virtues of Ganita Sutras is to be of Vedic order. For it, the sadkhas have to approach these Sutras, the Vedic way of Sankhiya Nishtha and Yoga Nishtha.
8. Ganita Sutras (and Ganita Upsutras) as a complete scripture of vedic order shall be availing Om (ॐ), the sole syllable Brahman formulation as the source reservoir of values and virtues for their sequential flow for manifestation of the formats for the whole range of the working rules for applied values of the pure values.
9. As this flow format of the values from the start with source reservoir leads to the Parnava domains, as such Om (ॐ), sole syllable Braham, as the starting point manifests at the end of the flow formats range, as Parnava प्रणव: as its synonym.
10. With it, the Ganita Sutras scripture begins with Om (ॐ) as source formulation and Parnava प्रणव: as the end fruit formulation.
11. Sadkhas fulfilled with the intensity of urge for enlightenment of the order the values and virtues of Ganita Sutras and their applications domains shall put themselves on the Vedic processing systems format path of Sankhiya Nishtha and Yoga Nishtha by permitting the transcending mind to be face to face with these processing processes and glimpse the phenomenon of artifices of numbers and dimensional frames running parallel to each other and being of complementary and supplementary values at every step
12. To begin with, the transcending mind shall be permitted to follow the transcendence path of Divya Ganga flow through the Om (ॐ) formulation starting with Bindu Sarovar ● of Brahman values where from the

transcendental values to flow as of seven streams which ahead shall be of three folds, three of the streams flowing towards east, the other three streams flowing towards west and the middle stream flowing ahead following the chariot of King Sagar.

13. This Divya Ganga flow path format is to be of Nav Braham values of nine streams reservoir within Bindu Sarovar (Point reservoir) as the first fold. These nine streams shall flow ahead as seven streams of Sapt Rishi Lok as the second fold of the Divya Ganga flow path. The split of this Sapt Rishi Lok, Divya Ganga flow into three folds as of three streams of eastern side, three streams of western side and the central stream would become the third fold of the Divya Ganga flow path. Finally, the central stream flow path shall be manifesting the flow path in the Bulok (Earth Sphere).
14. It is this central flow stream path in the Bulok (Earth Sphere) which is to manifest the values flow path for Vedic mathematics, science and technology as is manifesting as the scripture of Ganita Sutras. Within the Bulok (Earth Sphere) the existence phenomenon as Triloki/ 3-space/hyper cube 3/cube format, is to be the end fruit of the values manifestation format for the central stream of fourth fold of Divya Ganga flow format, shall be reaching the origin of 3-space/center of sphere/center of cube.
15. The features of this manifestation format of cube as such, is to be of the characteristics of the geometric envelope of the cube itself.
16. One may have a pause here and permit the transcending mind to have a fresh look at the set up of the geometric envelope of the cube.
17. One shall be finding as that there are eight corner points, six surfaces, twelve edges, and three dimensional axes together making out the manifested form of the geometric frame for the cube. These 29 features manifestation, as

such is to lead to the range of the Ganita Sutras and Ganita Upsutras.

18. Sam, the essence, as Samved format manifestation at the base of Samved Samhita as well shall be leading to the enlightenment as to the values of such range to be of precisely of the values of artifice 29 as of three folds, being of six values of Purva archive one value of the middle archive and 22 values of the Utter archive. Srimad Bhagwad Geeta for its fifth chapter is of the range of 29 shalokas, Srimad Durga Saptsati for its last, 13th chapter as well is accepting range precisely of 29 shalokas. It would be blissful exercise for the sadkhas to remain in prolonged sittings of trans and to be face to face with artifice 29 and to glimpse its values and virtues to comprehend and imbibe the values and virtues of the scripture of Ganita Sutras (and Ganita Upsutras) as of sixteen Sutras and thirteen Upsutras text.

## 2. Central Stream chase

19. Central stream chase is of Sunlight format. It is a chase for the seven stream flow from Nav Braham Reservoir. It leads to transcendental domains as such this chase path is the transcendental meditation path. It in a way helps transcends through the fourth fold of Divya Ganga flow through four fold Om (ॐ) formulation.
20. The seven streams flow from nine streams Brahman reservoir, as a step ahead leads to the five fold transcendental domain. The other way round, the processing in reverse order beginning with the central stream would lead to three fold stream and a step ahead would be the five fold stream domain of the values and virtues of five fold transcendental (5-space) domain.
21. This as such formats the transcending mind path as from above nine streams Brahman reservoir of seven streams flow transcending into the five fold transcendental (5-space) domain. The same from the other end of the single

stream, it would ascend to three fold stream leading to five fold transcendental (5-space) domain.

22. The sadkhas fulfilled with intensified urge to be face to face with this phenomenon of transcendence and ascendance approach, shall sit comfortably and permit the transcending mind to glimpse the transcendental (5-space) domain from both ends.
23. Sadkhas sitting comfortably and permitting his transcending mind to transcend and ascend in its natural way shall be glimpsing the transcendental (5-space) domain in both ways of transcendence as well as of ascendance.
24. The sadkhas shall permit the transcending mind to remain in prolonged glimpsing bliss ambrosia of the transcendental (5-space) domain at the middle of the Divya Ganga flow format.
25. Sadkhas shall remain in prolonged sittings of trans to permit the transcending mind to transcend deep within the transcendental (5-space) domain and to ascend from within it to be face to face with its inner folds.
26. The inner folds glimpsed while transcending along with the seven streams emanating from Brahman reservoir may be comprehending and imbibing one set of transcendental values and virtues while being through the inner folds of the transcendental (5-space) domain. Further while the transcending mind shall be ascending through the inner folds of the transcendental (5-space) domain, it shall be comprehending and imbibing another set of features of values and virtues of inner folds of the transcendental (5-space) domain.
27. One shall remain in prolonged sittings of trans as long as it would be blissful.
28. One shall go for trans sittings as many times as one feels blissful and till one's intensity of urge is not fulfilled with the transcendental enlightenment.

29. One may following the Sankhiya Nishtha format of artifices of numbers availed artifices 9, 7 and 5 for transcendence from above to be within the values format of artifice 5.
30. Further one may avail artifices 1, 3 and 5 to ascend to be along format of artifice 5.
31. One shall simultaneous transcend and ascend from above and from below along pair of artifices 9 and 7 on one side and artifices 1 and 3 on the other side.
32. Parallel to it one shall follow the Yoga Nishtha format of dimensional frames of 9-space, 7-space, 5-space and 3-space.
33. One may avail the formats of dimensional frames of 9-space and 7-space for transcendence into 5-space domain.
34. Further one shall avail the formats of dimensional frames of 1-space and 3-space for ascendance within 5-space.
35. One shall simultaneously avail geometric formats of 9-space and 7-space on the one side and 1-space and 3-space on the other side to transcend and ascend within 5-space.
36. One shall simultaneously avail the Sankhiya Nishtha and Yoga Nishtha formats for simultaneous transcendence and ascendance within the transcendental (5-space) domain along the Divya Ganga flow format through the artifices of Om (ॐ), sole syllable Brahman formulation.
37. One shall continue glimpsing the transcendental (5-space) domain through prolonged sittings of transcendence and ascendance following the central stream flow format which initially shall be leading up till the origin of 3-space/centre of cube. The Sama flow through the central stream shall be fulfilling the 3-space/cube and with it the solid dimensional order would manifest as the dimensional order of the transcendental space/domain/sky within space.
38. It is this attainment of the transcending mind of blissfully being along the Divya Ganga Flow path and to be within

the transcendental (5-space) domain which is to be of the enlightenment values and virtues which of their own shall be providing take off and riding the transcendental (5-space) carriers of the sunlight itself and thereby there being the reach to the source Brahman reservoir.

## 3. Glimpsing Origin of Transcendental (5-space) Domain

39. Sadkhas continue transcending and ascending along the central stream and to glimpse origin of the transcendental (5-space) domain.
40. This as such, as a result of continuous transcendence and ascendance shall be bringing the transcending mind face to face with the phenomenon of the central stream coordinating the origins and transforming itself as the origins coordination format for the transcendence and ascendance of the transcending mind itself.
41. It as such shall be bringing the transcending mind face to face with the structural set up of the origins of whole range. It is this format of structured origins whose transcendence from the previous state structural set up of origin to the following state of structural set up and back which shall be becoming the transcending mind mathematics, Science and technology. Ganita Sutras, as such manifest this mathematics, Science and technology of transcending mind of Vedic order.
42. This mathematics, Science and technology in its pure and applied forms provides formats for the transcending mind processing processes for the whole range of existence phenomenon, may it be as it is lively within human frame, of which, the mind is a integral part, and may it be within the Sun or otherwise.
43. Sadkhas fulfilled with intensity of urge for full range chase for the Ganita Sutras range shall permit the transcending mind to chase this range in the sequence and order of the Ganita Sutras organization itself. One shall sit comfortably and permit the transcending mind to chase this organization in its sequence and order by beginning with

the Ganita Sutra-1 itself and for its range as well beginning from its very first letter and sequentially to be through it letter by letter and syllable by syllable from the first syllable to the last syllable and from first letter to the last letter of the Sutra. One shall sit comfortably and permit the transcending mind to transcend through this sequential organization of the sutra.

44. One shall permit the transcending mind to transcend through the first letter ए/the sixth vowel.

45. One shall permit the transcending mind to remain in prolonged trans to transcend and ascend through this letter and this syllable to fully comprehend and to completely imbibe the form, frame, format, frequencies, values and virtues of this letter and this syllable before the transcending mind blissfully transcends ahead of it to the next letter 'क्' and the syllable 'क'.

46. One shall permit the transcending mind to continuously remain in deep prolonged sittings of trans as long as one is blissful with it. Between two sittings of trans, one may go through the scriptures and the lexographies and other Angas (shiksha, Vyakaran, Nirukt/Nighantu, Chandas, Jyotish and Kalp).

47. This step by step sequential validation of the experiential attainments of the transcending mind along the organization order of the Ganita Sutras in their external and internal sequential order of the scriptural text shall be ensuring required enlightenment of the range of the pure and applied values range of Ganita Sutras (and Ganita Upsutras).

## 4. 'क् बह्मा' /Ka Brahma

48. 'क्', the first consonant letter of Devnagri alphabet, is the creator letter. The Dictionaries, as such, associate it the meanings as 'Brahma' the creator. It is also, accordingly accepted as to be of the format of idol of Lord Brahma.

49. Lord Brahma, creator the supreme is the four head Lord and the presiding deity of 4-space, as such is associated with the number '4'.
50. The first letter of Ganita Sutra-1 is the sixth vowel and accordingly accepts association of artifices of number '6'.
51. Lord Vishnu, the Lord of self referral domains/6-space accepts association of artifices of number '6'.
52. Surya/Sun, Vishnu lok/Go lok/pursha as well accept artifices of number '6'.
53. With it the transition from first letter of Ganita Sutra-1, namely ','/sixth vowel/6-space/artifice 6 to the second letter of Ganita Sutra-1, namely 'क'/first varga consonant/Lord Brahma/4-space/Idol of Lord Brahma/ artifice 4 deserves to be comprehe-nded and imbibed.
54. Here one may have a pause and have a fresh look at the sequential transition steps of Divya Ganga flow, which begins with Brahman reservoir/Nav Braham/9-space/ artifice 9 and the same at the next step emanates 7 stream flow/sapt Rishi lok/7-space/artifice 7.
55. This, this way gives rise to the transition of features beginning with artifice 9 and reaching artifice 7.
56. The Ganita Sutras transition from first letter of Ganita Sutra-1 to the second letter of Ganita Sutra-1 is of the features of transition from artifice 6 to artifice 4.
57. One may have a pause here again and while having a fresh look at the pairs of artifices, (9, 7) and (6, 4) one may comprehend as that in both cases, here is going to be a jump over the middle artifices of the said pairs namely a jump over artifice 8 in case of the pair of the artifices (9, 7) and the jump over artifice 5 in case of the pair of artifices (6, 4).
58. This feature, as such shall be leading to the characteristics of symmetry which precisely is the rule of Upsutra-1 'आनुरूप्येण Anurupyena' whose simple English rendering comes to be as 'to follow the form as it is framed'.

59. Here again one may have a pause and a fresh look at all these features together and to comprehend and imbibe its values as to how the Ganita Sutra-1 sequentially leads to the rules of Ganita Upsutras.
60. The other way round it as well would help appreciate as to how the Ganita Sutras, here in particular, Ganita Upsutra-1, shall be helping reach at the transition features for the Ganita Sutras, here in particular Ganita Sutra-1, availing the original source, namely, the Brahman reservoir whose values and features give rise to the Divya Ganga flow.
61. For complete comprehension and for fully imbibing of the values and virtues of these features of the organization format of Ganita Sutras, one shall chase the vedic way of permitting the transcending mind of glimpsing it by being face to face to it.
62. One shall continuously remain in prolonged sittings of deep trans to comprehend and imbibe as to how the simultaneous reach from Ganita Sutra-1 to Ganita Upsutra-1 and back from Ganita Upsutra-1 to Ganita Sutra-1 is becoming possible.
63. For it begin with one may have a look at the working rules of Ganita Sutra-1 and Ganita Upsutra-1 themselves.
64. Working rule of Ganita sutra-1 'one more than before' and of Ganita Upsutra-1 'Symmetry/proportionately' while starting with Sutra-1 shall be providing a format for sequential progression of one step at a time. This while reaching Upsutra-1 shall be transiting into the format of sequential progressions of equal jumps at every step.
65. On the other way round the proportionate rule of Ganita Upsutra-1 shall be manifesting the rule of Sutra-1 with the features as that the one itself may be of any initial chosen measure value which shall be permitting one to be two, or three and so on. And also the other way round as that two to be one or three to be one and so on.

66. Here again one may have a pause and permit the transcending mind to be face to face with this two fold transition process simultaneously happening taking from Sutra-1 to Upsutra-1 and back from Upsutra-1 to sutra-1 while the rule of symmetry as there being a transition from Divya Ganga flow format to the format of Ganita Sutra-1 and Ganita Upsutra-1 together as integrated format. One may again have a pause here and have a fresh look at the format of interval/line/body of 1-space. The orientations of a line taking from one end to other end and back from second end to the first end, these shall be leading to the pair of formats of +1 and –1 parallel to +1 space and –1 space. Here is this pairing (+1, –1) there is a jump over artifice of number '0' /0-space/0-space body which is of the order and features of the transition from first step to the second step of Divya Ganga flow format as well as of transition from first letter of Ganita Sutra-1 to second letter of Ganita Sutra-1 and ahead from the working rule of Sutra-1 to the working rule of Upsutra-1.

## 5. 'क् बह्मा'/Ka Brahma to 'क् शिव' Ka Shiv'

67. Dictionaries give simultaneous meanings as 'क् बह्मा', 'क् शिव'.

68. Scriptures simultaneously accept Brahma as creator, and also Shiv as Creator.

69. Scriptures also enlightened as that Lord Brahma creates with the grace of Lord Shiv.

70. Lord Brahma, the four head lord is the presiding deity of 4-space with his idol being of the order and features of hyper cube 4, the representative regular body of 4-space.

71. Lord Shiv, the five head lord is the presiding deity of 5-spacwe with his idol being of the order and features of hyper cube 5, the representative regular body of 5-space within 4-space.

72. 5-space within 4-space plays the role of origin of 4-space.
73. The transition from domain to origin and back from origin to domain is parallel to the transcendence and ascendance process happening simultaneously. It is of the two fold sequential progression of ascending and descending order of the format of line/interval.
74. It as such is of the order and features of sequential progressions of the format of the line parallel to the working rule of Ganita Sutra-1.
75. One may have a pause here and permitting the transcending mind to chase this rule of sequential progression along format of a line as a track of a moving point.
76. With it follow the smooth transition from 4-space to 5-space which may help comprehend and imbibe the simultaneous meanings of values and virtues of the first varga consonant 'क' simultaneously accepting the meanings as "क् बह्मा', 'क् शिव".
77. One may have a pause here and permit the transcending mind to be face to face with this feature of letter "क्' simultaneously accepting the pair of the meanings 'क् बह्मा' as well as 'क् शिव' as being the rule of sequential transition.
78. This rule of sequential progression with a pair of end points being bridged by the transition feature being simultaneously available shall be like that of the expression range between the pair of end points of interval being designated as the length of the interval.
79. This, this way, shall be the inherent feature of the organization format of Ganita Sutra-1 being successful to attain continuity through discrete 's' values.
80. The attainment of continuity in terms of discrete values helps transit from the artifices of numbers as dimensional frames the continuous dimensional domains.

81. It is this transition from numbers to geometric formats and back from geometric formats to number values which becomes the postulates and axioms for the Sankhiya Nishtha and Yoga Nishtha; Sankhiya Nishtha presuming the existence of geometric format and availing artifices of numbers while on the other hand Yoga Nishtha presuming the existence of artifices of numbers and availing the geometric formats.

## 6. Ganita Upsutra-1

**UPSUTRA - I**

**आनुरूप्ये**

| 01 | 02 | 03 | 04 | 05 | 06 | 07 | 08 | 09 | 10 |
|---|---|---|---|---|---|---|---|---|---|
| आ | न् | उ | र् | ऊ | प् | य् | ए | ण् | अ |
| Total Letters | | | Vowels | | Nasels | | Consonents | | |
| | 10 | | 5 | | - | | 5 | | |

## 7. Initial Steps Outline

(*i*) Read the text of the Upsutra.

(*ii*) Pronounce the text Loudly.

(*iii*) Chase the text for its working rule

(*iv*) Think, meditate and transcend through the text of the Upsutra to have complete insight and for fully imbibing its values to perfect one's skills for applications of the working rule of the Upsutra.

(*v*) And to have the overview as follows

Text आनुरूप्येण

**Roman script Anurupyena**

**Text meanings**

'अनु' *means to follow, :i means the form and ':प्येण' means the form as framed. As such,* 'आनुरूप्येण' *means to follow the form as framed; symmetry; proportionately*

**Simple English Rendering: Symmety; Proportionately**

It is the source Upsutra. It is about Symmetry. The same is of the feature of "Proportionately." Literally it means "to follow the form as framed." With it, it has much wider meanings, in the context of Geometrical bodies and Geometrical formats.

This Upsutra when read along with any sutra, it adds and multiplies the meanings and operational domains of the concerned sutra. Illustratively, when Upsutra-1 is read alongwith Sutra-1, it shall be leading from whole numbers to fractions. Also it would make it applicable from Counting to tables. In fact, it shall be dealing with all Orders from a Sequences, series and sets of single and multiple operators, and so on.

## 8. Ganita Upsutra-2:

**UPSUTRA - II शिष्यते शेषसंज्ञ:**

| 01 | 02 | 03 | 04 | 05 | 06 | 07 | 08 | 09 | 10 |
|---|---|---|---|---|---|---|---|---|---|
| श् | इ | ष् | य् | अ | त् | ए | श् | ए | ष् |

| 11 | 12 | 13 | 14 | 15 | 16 | 17 | 18 |
|---|---|---|---|---|---|---|---|
| अ | स् | अ | | ज् | ञ् | अ | : |

| Total Letters | Vowels | Nasels | Consonents |
|---|---|---|---|
| 18 | 7 | 2 | 9 |

## 9. Initial Steps Outline

(*i*) Read the text of the Upsutra.

(*ii*) Pronounce the text Loudly.

(*iii*) Chase the text for its working rule

(*iv*) Think, meditate and transcend through the text of the Upsutra to have complete insight and for fully imbibing its values to perfect one's skills for applications of the working rule of the Upsutra.

(*v*) And to have the overview as follows

Text शिष्यते शेषसंज्ञ:

Roman script *Sisyate Sesasamjnah*

Text meanings chase *शिष्यते means that remains and शेष means the remainder and संज्ञः means noun. As such शिष्यते शेषसंज्ञः means that 'name* (*noun*) *of 'that remains' is 'remainder'*

## Simple English Rendering: That remains is remainder

As a step, ahead of Upsutra-1, and as a transition and break through there from, it shall be shifting from Symmetrical domains to asymmetrical domains. Literally the Upsutra means as that which remains is designated as the remainder. It in a way, takes care of the remainders on division. In the context of Place value system, it shall be focusing upon the Unit Place values.

However, when the Upsutra-2 is read alongwith Sutra-2, it shall be providing transition from ten place value system to other place value system. This way new dimensions would get added to the applied values of Sutra-2 range.

## 10. Ganita Upsutra-3

**UPSUTRA - III**

आद्यमाद्येनान्त्यमन्त्येन

| 01 | 02 | 03 | 04 | 05 | 06 | 07 | 08 | 09 | 10 |
|---|---|---|---|---|---|---|---|---|---|
| आ | द् | य् | अ | म् | आ | द् | य् | ए | न् |
| 11 | 12 | 13 | 14 | 15 | 16 | 17 | 18 | 19 | 20 |
| आ | न् | त् | य् | अ | म् | अ | न् | त् | य् |
| 21 | 22 | 23 | | | | | | | |
| ए | न् | अ | | | | | | | |

| Total Letters | Vowels | Nasels | Consonents |
|---|---|---|---|
| 23 | 9 | - | 14 |

## 11. Initial steps outline

(*vi*) Read the text of the Upsutra.

(*vii*) Pronounce the text Loudly.

(*viii*) Chase the text for its working rule

(*ix*) Think, meditate and transcend through the text of the Upsutra to have complete insight and for fully imbibing its values to perfect one's skills for applications of the working rule of the Upsutra.

(*x*) And to have the overview as follows

आद्यमाद्येन अन्त्यमन्त्येन *Adyamadyen Antyamantyena/*

First with first Last with last

आदि: means the beginning/first. अन्त: means the end/last.

As such आद्यमाद्येन means the first with the first and अन्त्यमन्त्येन means last with the last

This is a transition step ahead of the already transitions attained uptil Upsutra-2. The original transition from Upsutra-1 to Upsutra-2 has been of reaching from Symmetry domain to asymmetry domain. However here, at the present phase and stage of Upsutra-3, the applications of dealing with asymmetrical situations are being taken care of. The beginning points, and end points of two domains are being tagged for the simultaneous coverage of two domains, may or may not be those be equal. It is a big step. It adds to the new set ups of the formats.

## 12. Ganita Upsutra-4

**UPSUTRA - IV**

**केवलै: ससकं गुण्यात्**

| 01 | 02 | 03 | 04 | 05 | 06 | 07 | 08 | 09 | 10 |
|---|---|---|---|---|---|---|---|---|---|
| क् | ए | व् | अ | ल् | ऐ | : | स् | अ् | प् |
| 11 | 12 | 13 | 14 | 15 | 16 | 17 | 18 | 19 | 20 |
| त् | अ | क् | अ | ं | ग् | अ | ण् | य् | आ |
| 21 | | | | | | | | | |
| त् | | | | | | | | | |

| Total Letters | Vowels | Nasels | Consonents |
|---|---|---|---|
| 21 | 8 | 2 | 11 |

## 13. Initial Steps Outline

(*i*) Read the text of the Upsutra.

(*ii*) Pronounce the text Loudly.

(*iii*) Chase the text for its working rule

(*iv*) Think, meditate and transcend through the text of the Upsutra to have complete insight and for fully imbibing its values to perfect one's skills for applications of the working rule of the Upsutra.

(*v*) And to have the overview as follows

केवलै: सप्तकं गुण्यात्/ *Kevalaih Saptakam Gunyat/*

Only Seven as multiplicand

केवलै: means the 'only'ण सप्तकं means 'seven'. And 'गुण्यात्' means the multiplicand.

As such केवलै: सप्तकं गुण्यात् means only seven as multiplicand.

It takes a step ahead. As the Upsutra literally means as that "features comparisons can be only uptil seven", it shall be about the upper limit being that of artifice seven and parallel to it the seven space. It as such focuses upon the artifice seven, being the biggest Prime Numeral for the ten place value system. Further as a Geometrical property, it is focusing upon the feature as that it is only uptil Hyper circle seven that there is to be an increase. It as such, is a big range beyond which the Mathematics is to be different. It in a way, is the limit of the linear order because of which linear order three space excepts range of seven Geometries only. Beyond that, is to come into play is spatial order.

## 14. Ganita Upsutra-5

**UPSUTRA - V**

**वेष्टनम्**

| 01 | 02 | 03 | 04 | 05 | 06 | 07 | 08 |
|---|---|---|---|---|---|---|---|
| व् | ए | ष | ट् | अ | न् | अ | म् |

| Total Letters | Vowels | Nasels | Consonents |
|---|---|---|---|
| 8 | 3 | - | 5 |

## 15. Initial Steps Outline

(*vi*) Read the text of the Upsutra.

(*vii*) Pronounce the text Loudly.

(*viii*) Chase the text for its working rule

(*ix*) Think, meditate and transcend through the text of the Upsutra to have complete insight and for fully imbibing its values to perfect one's skills for applications of the working rule of the Upsutra.

(*x*) And to have the overview as follows

It is about Oscillators. Which is to take care of Cycles within Cycles. It is a transitory phase and stage from the linear order logic. It take cares of Divisibility. It is about factors. It would help chase the structural set-ups of domains. In fact, it is here that the Vedic Mathematics starts marking its presence for the Higher Dimensional reality set-ups.

## 16. Ganita Upsutra-6

**UPSUTRA - VI**

यावदूनं तावदूनम्

| 01 | 02 | 03 | 04 | 05 | 06 | 07 | 08 | 09 | 10 |
|---|---|---|---|---|---|---|---|---|---|
| य् | आ | व् | अ | द् | ऊ | न् | अ | | त् |
| 11 | 12 | 13 | 14 | 15 | 16 | 17 | 18 | | |
| आ | व् | अ | द् | ऊ | न् | अ | म् | | |

| Total Letters | Vowels | Nasels | Consonents |
|---|---|---|---|
| 18 | 8 | 1 | 9 |

## 17. Initial Steps Outline

(*i*) Read the text of the Upsutra.

(*ii*) Pronounce the text Loudly.

(*iii*) Chase the text for its working rule

(*iv*) Think, meditate and transcend through the text of the Upsutra to have complete insight and for fully imbibing its values to perfect one's skills for applications of the working rule of the Upsutra.

(*v*) And to have the overview as follows

यावदूनं तावदूनम्/ *Yavadunam Tavadunam*/That twice This twice

It is in Continuity of the transitory features being taken care of with the phase and stage of Upsutra-5. Here the Deficiencies in multiplied forms are taken care of. It takes to the inner folds of the structural domains.

## 18. Ganita Upsutra-7

**UPSUTRA - VII**

**यावदूनं तावदूनीकृत्य वर्गं च योजयेत्**

| 01 | 02 | 03 | 04 | 05 | 06 | 07 | 08 | 09 | 10 |
|---|---|---|---|---|---|---|---|---|---|
| य् | आ | व् | अ | द् | ऊ | न् | अ | | त् |
| 11 | 12 | 13 | 14 | 15 | 16 | 17 | 18 | 19 | 20 |
| आ | व् | अ | द् | ऊ | न् | ई | क् | ऋ | त् |
| 21 | 22 | 23 | 24 | 25 | 26 | 27 | 28 | 29 | 30 |
| य् | अ | व् | अ | र् | ग् | अ | | च् | अ |
| 31 | 32 | 33 | 34 | 35 | 36 | 37 | | | |
| य् | ओ | ज् | अ | य् | ए | त् | | | |

| Total Letters | Vowels | Nasels | Consonents |
|---|---|---|---|
| 37 | 16 | 2 | 19 |

यावदूनम् तावदूनीकृत्य वर्गं च योजयेत्/ *Yavadunam Tavadunikrtya Varganca Yogayet*/That twice This twice Square and add

**It attains transitions from linear order to spatial order. It takes to the features of a square.**

## 19. Ganita Upsutra-8

**UPSUTRA - VIII**

अन्त्ययोर्दशकेऽपि

| 01 | 02 | 03 | 04 | 05 | 06 | 07 | 08 | 09 | 10 |
|---|---|---|---|---|---|---|---|---|---|
| अ | न् | त् | य् | अ | य् | ओ | द् | अ | र् |
| 11 | 12 | 13 | 14 | 15 | 16 | | | | |
| श् | अ | क् | ए | प् | इ | | | | |

| Total Letters | Vowels | Nasels | Consonents |
|---|---|---|---|
| 16 | 7 | - | 8 |

अन्त्पयोर्दशकेऽपि/ *Antyayordasake'pi*/Ends to sum as ten
The transitions attained with the above step of Upsutra-7 leading from linear order to spatial order, it sequentially is taken to be carried forward uptil artifice ten. It is in respect of the end parts. It as such is concerning the boundaries. The boundary as of ten components, that is that of Hypercube-5 shall be the attainment format. Here it may be relevant to note as that this parallel sequential steps of Upsutras. In fact are to be chased alongwith the sequential order of the Sutras.

## 20. Ganita Upsutra-9

**UPSUTRA - IX**

अन्त्ययोरेव

| 01 | 02 | 03 | 04 | 05 | 06 | 07 | 08 | 09 | 10 |
|---|---|---|---|---|---|---|---|---|---|
| अ | न् | त् | य् | अ | य् | ओ | र् | ए | व् |
| 11 | | | | | | | | | |
| अ | | | | | | | | | |

| Total Letters | Vowels | Nasels | Consonents |
|---|---|---|---|
| 11 | 5 | - | 6 |

अन्तययोरेव/ *Antyayoreva*/Ends to be in ratio
It takes the progression a step ahead. It as such becomes the General chase for the boundaries of the Hyper-cubes 5 onwards.

## 21. Ganita Upsutra-10

**UPSUTRA - X**

समुच्चयगुणित:

| 01 | 02 | 03 | 04 | 05 | 06 | 07 | 08 | 09 | 10 |
|---|---|---|---|---|---|---|---|---|---|
| स् | अ | म् | उ | च् | च् | अ | य् | अ | ग् |
| 11 | 12 | 13 | 14 | 15 | 16 | | | | |
| उ | ण् | इ | त् | अ | : | | | | |

| Total Letters | Vowels | Nasels | Consonents |
|---|---|---|---|
| 16 | 7 | 1 | 8 |

समुच्चयगुणित:/ *Samuccayagunitah*/Samuchya as product
It is all about the sets with structural operators transiting from the manifested domains of such structural set-ups. It in a way, is

to transcend from the manifestation formats as that of the Hyper-cubes. This, this way is a very big step forward. As here, the Mathematics is going transcendental.

## 22. Ganita Upsutra-11

**UPSUTRA - XI**

लोपनस्थापनाभ्याम्

| 01 | 02 | 03 | 04 | 05 | 06 | 07 | 08 | 09 | 10 |
|---|---|---|---|---|---|---|---|---|---|
| ल् | ओ | प् | अ | न् | अ | स् | थ् | आ | प् |
| 11 | 12 | 13 | 14 | 15 | 16 | 17 | | | |
| अ | न् | आ | भ् | य् | आ | ण् | | | |

| Total Letters | Vowels | Nasels | Consonents |
|---|---|---|---|
| 17 | 7 | - | 10 |

लोपन स्थापनाभ्याम्/ *Lopana Sthapananabhyam*/That missing to be established

It is the feature of transcedental domains re-manifesting. It is of the features of un-manifest, remanifesting by ascendance from the origin fold into the domain fold.

## 23. Ganita Upsutra 12

**UPSUTRA - XII**

विलोकनम्

| 01 | 02 | 03 | 04 | 05 | 06 | 07 | 08 | 09 |
|---|---|---|---|---|---|---|---|---|
| व् | इ | ल् | ओ | क् | अ | न् | अ | म् |

| Total Letters | Vowels | Nasels | Consonents |
|---|---|---|---|
| 9 | 4 | - | 5 |

विलोकनम्/ *Vilokanam*/By observation

It is about the capabilities to glimps and to be face to face with the re-manifested transidental domains phenomena.

## 24. Ganita Upsutra 13

**UPSUTRA - XIII**

गुणितसमुच्चय: समुच्चयगुणित

| 01 | 02 | 03 | 04 | 05 | 06 | 07 | 08 | 09 | 10 |
|---|---|---|---|---|---|---|---|---|---|
| ग् | उ | ण् | इ | त् | अ | स् | अ | म् | उ |
| 11 | 12 | 13 | 14 | 15 | 16 | 17 | 18 | 19 | 20 |
| च् | च् | अ | य् | अ | : | स् | अ | म् | उ |
| 21 | 22 | 23 | 24 | 25 | 26 | 27 | 28 | 29 | 30 |
| च् | च् | अ | य् | अ | ग् | उ | ण् | इ | त् |
| 31 | 32 | | | | | | | | |
| अ | : | | | | | | | | |

| Total Letters | Vowels | Nasels | Consonents |
|---|---|---|---|
| 32 | 14 | 2 | 16 |

गुणितसमुच्चय: समुच्चयगुणित: / *GunitaSamuccaya Samuccayagunitah/* Product samuchya Samuchya Product.

It is the culmination of the renewing cyclic features of the transcidental domains which firstly strips off the manifested domains of manifestation features and puts into a transcendental set/ domain, and then again under a transcendental renewal process remenifest the same as renewed transcendental domain.

These progression features process of Upsutras beginning with the Symmetry and reaching uptil the renewed transcidental domains deserves to be chased as such to imbibe the values and virtues range of Ganita Upsutras.

❑ ❑ ❑